Their words
my thoughts

Oxford University Press 1981

Contents

✠ In praise of God   6

⌣ Jesus Christ   44

◉ The way of life   114

⊗ The world around us   158

Index   188

Acknowledgements   191

Oxford University Press, Walton Street, Oxford OX2 6DP

London   Glasgow   New York   Toronto
Delhi   Bombay   Calcutta   Madras   Karachi
Kuala Lumpur   Singapore   Hong Kong   Tokyo
Nairobi   Dar es Salaam   Cape Town   Salisbury
Melbourne   Wellington

and associate companies in
Beirut   Berlin   Ibadan   Mexico City

Cover photograph courtesy of Kodak Ltd.

Typeset by Tradespools Ltd., Frome, Somerset
Music set by Rowland Phototypesetting, Bury St. Edmunds, Suffolk
Printed in Hong Kong

Introduction

In this book you will find hymns and songs, poems, thoughts and pictures. These can help you to think about yourself, others and the world around you.

The words, music and pictures of other people help you to understand what they have thought and felt about these things and about God and His Son, Jesus Christ.

It is not only their words which are important but also what they mean to you as you think about them. That is the reason for calling this book *Their Words, My Thoughts.*

You will use it many times during assemblies, but you can also use it with your friends in class or when you are by yourself. We hope this book will help you to express your thoughts in words, music and pictures, and to share these with others.

The ten commandments

I am the Lord your God.

You shall have no other god before me.

You shall not make a carved image for
yourself.

You shall not bow down to
them or worship them.

You shall not make wrong use of the name
of the Lord your God.

Remember to keep the sabbath holy.

Honour your father and your mother.

You shall not commit murder.

You shall not commit adultery.

You shall not steal.

You shall not give false evidence
against your neighbour.

You shall not covet your neighbour's
house – or anything that belongs to him.

Exodus 20 vv 2–17

Summary of commandments

You shall love the Lord your God with all your heart,
and with all your soul, and with all your strength,
and with all your mind; and you shall love your
neighbour as yourself.

Luke 10 v 27

The Lord's Prayer

John Whitworth, b.1921

Our Father, who art in heav'n, hallow'd be thy name.

Thy King-dom come, thy will be done On earth, as it is in

heav'n. Give us this day our daily bread, And for-

give us our tres-pas-ses As we forgive those who tres-pass a-

gainst us. And lead us not in-to temp-ta-tion, But deliver us from

e-vil. For thine is the king-dom, the pow'r and the

glo-ry. For e-ver and e-ver. A - men.

★ *Alternative ending for short Lord's Prayer:*

e - vil. A - men.

✠ 1 All creatures of our God and King

William Henry Draper, 1855–1933, based on St Francis of Assisi
Geistliche Kirchengesang (Cologne), 1623, arranged and harmonized by
Ralph Vaughan Williams, 1872–1958

EASTER ALLELUIA

1. All crea-tures of our God and King, Lift
up your voice and with us sing Al - le -
lu - ia, Al-le - lu - ia! Thou bur-ning sun with gol-den
beam, Thou sil-ver moon with soft - er gleam:

Refrain:
O praise him, O praise him, Al - le-
lu - ia, Al-le - lu - ia, Al-le - lu - ia!

2 Thou rushing wind that art so strong,
　Ye clouds that sail in heaven along,
　　O praise him, alleluia!
　Thou rising morn, in praise rejoice,
　Ye lights of evening, find a voice:
　Refrain:

3 Thou flowing water, pure and clear,
  Make music for thy Lord to hear,
    Alleluia, alleluia!
  Thou fire so masterful and bright,
  That givest man both warmth and light:
  Refrain:

4 Dear mother earth, who day by day
  Unfoldest blessings on our way,
    O praise him, alleluia!
  The flowers and fruits that in thee grow
  Let them his glory also show:
  Refrain:

The first morning of all

Then God reached out and took the light in his hands,
And God rolled the light around in his hands,
Until he made the sun;
And he set that sun a-blazing in the heavens.
And the light that was left from making the sun
God gathered it up into a shining ball
And flung against the darkness,
Spangling the night with the moon and stars.
Then down between the darkness and the light
He hurled the world;
And God said: 'That's good!'
Then the green grass sprouted,
And the little red flowers blossomed,
The pine-tree pointed his finger to the sky,
And the oak spread out his arms:
The lakes cuddled down in the hollows of the ground,
And the rivers ran to the sea;
And God smiled again,
And the rainbow appeared,
And curled itself around his shoulder.
Then God raised his arm and waved his hand
Over the sea and over the land,
And he said: 'Bring forth! Bring forth!'
And quicker than God could drop his hand,
Fishes and fowls
And beasts and birds
Swam the rivers and the seas,
Roamed the forest and the woods,
And split the air with their wings.
And God said: 'That's good!'

J. W. Johnson

✖ 2 Alleluia

Round in 3 parts

Con brio ♩ = 80–88

William Boyce, c.1710–79

|1|

Al - le - lu - ia, Al - le -

lu ia.

|2|

Al - le - lu - ia, Al - le - lu - ia, Al - le - lu -

ia, Al - le - lu - ia, Al - le - lu - ia.

|3|

Al - le - lu - ia, Al - le - lu - ia, Al -

le - lu - ia, Al - le - lu - ia.

Prayer

People ought to praise you, God of earth and heaven.
All of us ought to praise you.
You are always there, never growing old, fresh as each new day:
You were in Jesus, showing your love by his death, and by his
    resurrection giving us hope of living with you for ever:
You bring light and life to the world by your Holy Spirit,
    making every moment your moment, and every day your day
        of coming to the rescue.
To God the Father, God the Son, God the Holy Spirit let all
    the world give praise, today, and every day,
        and for ever and ever. *Amen*

Symbols of God

Christians do not try to
show what God is like in
pictures, but you may see
symbols in church
windows or carvings
which represent or stand
for God.

The all-seeing eye is an
old symbol of God. It
means that he knows
everything that goes on
in his world.

The three sides of the
triangle or the three
leaflets of the trefoil, or
clover-leaf shape, show
God as the Trinity, which
means:
God the Father who is
God the loving Creator.
God the Son who is God
showing what he is like in
Jesus.
God the Holy Spirit who
is God at work in the
world.

The hand of God is
another symbol of God. It
means that he is always
doing things in his world.
The hand is often shown
raised in blessing. It is
sometimes shown
holding, or looking after,
his people.

✠ 3 All people that on earth do dwell

W. Kethe, Daye's Psalter, 1560–1
Melody from Genevan Psalter, 1551

OLD HUNDREDTH

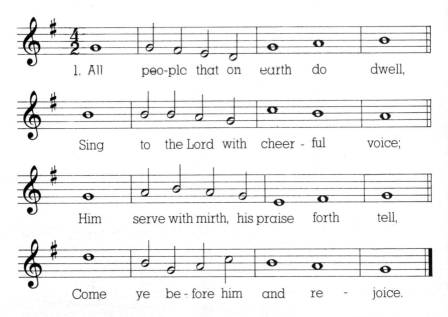

1. All peo-ple that on earth do dwell,
Sing to the Lord with cheer - ful voice;
Him serve with mirth, his praise forth tell,
Come ye be - fore him and re - joice.

2 The Lord, ye know, is God indeed,
  Without our aid he did us make;
We are his folk, he doth us feed,
  And for his sheep he doth us take.

3 O enter then his gates with praise;
  Approach with joy his courts unto,
Praise, laud, and bless his name always,
  For it is seemly so to do.

4 For why? the Lord our God is good:
  His mercy is for ever sure,
His truth at all times firmly stood,
  And shall from age to age endure.

5 To Father, Son, and Holy Ghost,
  The God whom heaven and earth adore,
From men and from the angel-host
  Be praise and glory evermore.

Praise the Lord!
Praise God in his Temple;
   praise him for his power in heaven!
Praise him for his mighty deeds:
   praise him according to his excellent greatness!

Praise him with the sound of the trumpet;
   praise him with the lute and harp!
Praise him with drums and dancing;
   praise him with strings and pipes!
Praise him with sounding cymbals;
   praise him with loud clashing cymbals!
Let everything that breathes praise the Lord!
Praise the Lord!

Psalm 150

Know that the Lord is God!
It is he that made us, and we are his;
we are his people, and the sheep of his pasture.

Psalm 100 v 3

True religion is to love all creatures, whether great or small, as
God has loved them.

Hindu saying

11

✠ 4 Everything changes but God

From *Goethe, Percy Dearmer*, 1867–1936
Martin Shaw, 1875–1958

GUN HILL

1. E-very-thing chan-ges, But God chan-ges not; The
power ne-ver chan-ges That lies in his thought:

Refrain:

*Splen-dours three, from God pro-cee-ding, May we e-ver
love them true, Good-ness, Truth and Beau-ty hee-ding
E-very day, in all we do.*

2 Truth never changes,
   And Beauty's her dress,
   And Good never changes,
   Which those two express:
   Refrain:

3 Perfect together
   And lovely apart,
   These three cannot wither;
   They spring from God's heart:
   Refrain:

4 Some things are screening
   God's glory below;
   But this is the meaning
   Of all that we know:
   Refrain:

12

Pied beauty

Glory be to God for dappled things –
For skies of couple-colour as a brinded cow;
For rose-moles all in stipple upon trout that swim;
Fresh-firecoal chestnut-falls; finches' wings;
Landscape plotted and pieced – fold, fallow, and plough;
And all trades, their gear and tackle and trim.
All things counter, original, spare, strange;
Whatever is fickle, freckled (who knows how?)
With swift, slow; sweet, sour; adazzle, dim;
He fathers-forth whose beauty is past change:
Praise him.

Gerard Manley Hopkins

Prayer

God, our loving Father, thank you that you never change. You are as strong and wise and loving as when you made the world. Thank you that nothing can ever happen that will make you alter. You are the one true God and Maker of all. We worship you and praise your holy name.
Amen

✠ 5 Father, we adore you

Round in 3 parts

Terrye Coelho
Terrye Coelho

Slowly, sustained

1. Fa - ther
2. Je - sus } we a-dore you; lay our lives be-
3. Spi - rit

fore you. How we love you.

Prayer

Day by day,
Dear Lord, of thee three things I pray;
To see thee more clearly,
Love thee more dearly,
Follow thee more nearly,
Day by day. *Amen*

Father, we thank you 6

ALL KINDS OF LIGHT

Caryl Micklem
Caryl Micklem, b.1925

1. Fa-ther, we thank you. For the light that shines all the day; For the bright sky you have gi - ven, Most like your hea-ven; Father, we thank you.

2 Father, we thank you.
  For the lamps that lighten the way;
  For human skill's exploration
  Of your creation;
  Father, we thank you.

3 Father, we thank you.
  For the friends who brighten our play;
  For your command to call others
  Sisters and brothers;
  Father, we thank you.

4 Father, we thank you.
  For your love in Jesus today,
  Giving us hope for tomorrow
  Through joy and sorrow;
  Father, we thank you.

✠ 7 Gloria in excelsis Deo

Round in 3 parts

Christopher le Fleming
Christopher le Fleming, b.1908

To be sung like a peal of bells ♩ = 144

Glo - ri-a, glo - ri-a, glo - ri-a in ex-cel-sis

De-o, in ex-cel-sis De - o, Glo-ri a, glo-ri - a,

in ex - cel - sis De - o, in ex-cel-sis De - o.

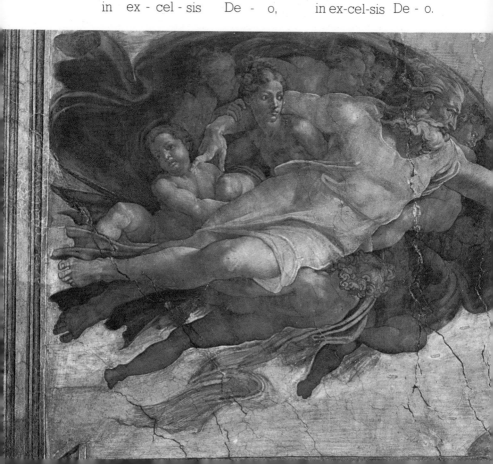

Rejoice in the Lord always: and again I say, rejoice.

Philippians 4 v 4

Prayer

Almighty God, whose glory the heavens are telling, the earth thy power and the sea thy might, and whose greatness all feeling and thinking creatures everywhere herald; to thee belongeth glory, honour, might, greatness and magnificence now and for ever. *Amen*

From the Liturgy of St James, 2nd century

✠ 8 For all the love

L. J. Egerton Smith, 1879–1958
John Whitworth, b.1921

1. For all the love that from our earliest days Has glad-den'd life and guar-ded all our ways, We bring Thee, Lord, our song of grate-ful praise, Hal-le-lujah, Halle-lu-jah, Halle-lu-jah, Halle-lujah, Halle-lujah, Halle-lu-jah

2. For -lu- jah, Hal-le -lu-jah, Hal-le-lu-jah, Hal-le-lu-jah, Hal-le-lu - jah.

2 For all the joy that childhood's days have brought,
For healthful lives and purity of thought,
For life's deep meaning to our spirits taught,
Hallelujah! Hallelujah!

3 For all the hope that sheds its glorious ray
Along the dark and unknown future way,
And lights the path to God's eternal day,
Hallelujah! Hallelujah!

4 For Christ the Lord, our Saviour and our friend,
Upon whose love and truth our souls depend,
Our hope, our strength, our joy that knows no end,
Hallelujah! Hallelujah!

Prayer

Lord, make us the instruments of your peace.
Where there is hatred, may we bring love;
Where there is malice, may we bring pardon;
Where there is discord, may we bring harmony;
Where there is error, may we bring truth;
Where there is doubt, may we bring faith;
Where there is despair, may we bring hope;
Where there is darkness, may we bring light;
Where there is sadness, may we bring joy. *Amen*

St Francis of Assisi, 1182–1226

✠ 9 Glory to thee, my God, this night

TALLIS' CANON
Round in 4 parts

Bishop T. Ken, 1637–1711
Thomas Tallis, c.1510–85

1. Glo - ry to thee, my God, this night For
all the bles-sings of the light; Keep me, O keep me,
King of Kings, Be - neath thy own al - migh-ty wings.

* *This canon may also be performed by 8 voices, entering at 4 notes' distance.*

2 Forgive me, Lord, for thy dear Son,
  The ill that I this day have done,
  That with the world, myself, and thee,
  I, ere I sleep, at peace may be.

3 Praise God, from whom all blessings flow;
  Praise him, all creatures here below;
  Praise him above, ye heavenly host;
  Praise Father, Son, and Holy Ghost.

The Lord bless you and keep you;
the Lord make his face to shine upon you
and be gracious unto you;
The Lord lift up his countenance upon you
and give you peace.

Numbers 6 vv 24–26

Prayer

O God, our Father, please forgive me for the wrongs I have done, for
bad tempers and angry words, for being greedy and wanting the best
only for myself, for making other people unhappy; forgive me, O God.
Amen

Talking to God

I talk to God about my hat,
And where I grazed my knee,
He likes to hear about our cat
Or what I had for tea.
My bicycle and where I went
My little brother's cough,
The lovely present Grandma sent,
The button that came off.
I tell him these and let him see
When I am sad or glad,
I talk to him about my friends
And how they make me mad,
He's never tired of listening
To things I want to tell
Because he is my Father
And he loves me very well.

Sister Monica Mary

✠ 10 God is love; his the care

✠ P. Dearmer, 1867–1936
Melody from *Piae Cantiones*, 1582,
arranged by Gustav Holst, 1874–1934

THEODORIC

1. God is love; his the care,

Ten-ding each e-very-where. God is love all is there.

Je-sus came to show him, That man-kind might know him:

Refrain:

Sing a-loud, loud, loud. Sing a-loud, loud, loud.

God is good. God is truth. God is beau-ty. Praise him.

Chime Bars

9

Refrain:

2 None can see God above;
  All have here man to love;
  Thus may we Godward move
  Finding him in others,
  Holding all men brothers:
  Refrain:

3 To our Lord praise we sing –
  Light and life, friend and king,
  Coming down love to bring,
  Pattern for our duty,
  Showing God in beauty:
  Refrain:

Maybe a stranger is a friend
you haven't met yet.

Traditional saying

Show love to all creatures,
and thou wilt be happy;
for when thou lovest all things,
thou lovest the Lord,
for he is all in all.

Tulsi Das, 1532–1623 (Hindu)

✳ 11 God is love: let heaven adore him

ABBOT'S LEIGH

Timothy Rees
Cyril V. Taylor, b.1907

God is love: let heav'n a - dore him;
God is love: let earth re -joice; Let cre - a-tion
sing be-fore him; And e - xalt him with one voice.
He who laid the earth's foun - da-tion, He who
spread the heav'ns a - bove, He who breathes through
all cre - a-tion, He is love, e - ter-nal love.

Prayer

Praise be to God, Lord of the worlds! The compassionate, the
merciful! King on the day of reckoning! Thee only do we worship,
and to thee do we cry for help. *Amen*

Koran

Listen for three kinds of word: the merciful word, the singing word
and the good word. These are of God and if you hear such words, you
hear God speaking.

A Gaelic saying

The names and titles of God

Long ago the Jews spoke a language called Hebrew, which was written without vowels. They wrote the name of God as YHWH, which we say as YAHWEH. It is sometimes called Jehovah. Jews think that this name is so special and holy that they never say it aloud. Instead they call God ADONAI, which means LORD.

In the Bible God is given many different titles.
He is called: the Almighty, or all-powerful God,
the Eternal, or everlasting God,
the Creator, or maker of all things,
the Holy One, or the one who is set apart or quite different from anyone or anything in his creation.
Jesus gave him another title: Father. He said that we can call God our Father, as he loves all people as a good father loves his children.

✠ 12 God of concrete

Richard G. Jones, b.1926
John Whitworth, b.1921

LEICESTER

1. God of con-crete, God of steel, God of pi-ston
and of wheel, God of py-lon, God of steam,
God of gir-der and of beam, God of a-tom, God of mine,
All the world of power is thine.

2 Lord of cable, Lord of rail,
Lord of motorway and mail,
Lord of rocket, Lord of flight,
Lord of soaring satellite,
Lord of lightning's livid line,
All the world of speed is thine.

3 Lord of science, Lord of art,
God of map and graph and chart,
Lord of physics and research,
Word of Bible, faith of Church,
Lord of sequence and design,
All the world of truth is thine.

Prayer

O Father of wisdom, we thank you for the gift of science, and for all who work to make new discoveries for the benefit of mankind. We pray you to bless them and to help men everywhere to use this knowledge in your service. *Amen*

I can't abide to see men throw away their tools i' that way, the minute the clock begins to strike, as if they took no pleasure in their work, and was afraid o' doing a stroke too much. I hate to see a man's arms drop down as if he was shot, before the clock's fairly struck, just as if he'd never a bit o' pride and delight in's work. The very grindstone 'ull go on turning a bit after you loose it.

George Eliot, 1819–80

✠ 13 God who made the earth

Sarah Betts Rhodes, 1830–90
H. von Muller, 1859–1938

SOMMERLIED

1. God who made the earth, The air, the sky, the sea,
Who gave the light its birth, Ca-reth for me.

2 God, who made the grass,
  The flower, the fruit, the tree,
The day and night to pass,
  Careth for me.

3 God who made the sun,
  The moon, the stars, is he
Who, when the day is done,
  Careth for me.

Prayer

Dear God, be good to me;
the sea is so wide,
and my boat is so small. *Amen*

Prayer of the Breton fishermen

The Lord is good to all men,
and his tender care rests upon all his creatures.

Psalm 145 v 9

Before the mountains were brought forth,
or ever thou hadst formed the earth and the world,
from everlasting to everlasting
thou art God.

Psalm 90 v 2

✠ 14 He's got the whole wide world

Negro spiritual,
arranged by D. J. Crawshaw

2 He's got ev'rybody here, in his hands,
He's got ev'rybody here, in his hands,
He's got ev'rybody here, in his hands,
He's got the whole world in his hands.

3 He's got the tiny little baby, in his hands,
He's got the tiny little baby, in his hands,
He's got the tiny little baby, in his hands,
He's got the whole world in his hands.

4 He's got you and me brother, in his hands,
He's got you and me sister, in his hands,
He's got you and me brother, in his hands,
He's got the whole world in his hands.

The tale of a garden: God creates man

Then God sat down –
On the side of a hill where he could think;
By a deep wide river he sat down;
With his head in his hands,
God thought and thought,
Till he thought: 'I'll make a man.'

Up from the bed of the river
God scooped the clay;
And by the bank of the river
He kneeled him down:
And there the great God Almighty,
Who lit the sun and fixed it in the sky,
Who flung the stars to the most far corner of the night,
Who rounded the earth in the middle of his hand:
This Great God,
Like a mammy, bending over his baby,
Kneeled down in the dust
Toiling over the lump of clay
Till he shaped it in his own image.

Then into it he blew the breath of life,
And man became a living soul. Amen

J. W. Johnson

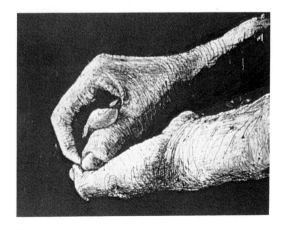

✠ 15 Let all the world

George Herbert, 1593–1633
Basil Harwood, 1859–1949,

LUCKINGTON

1. Let all the world in e-very cor-ner sing, My God and
King! The heav'ns are not too high, His praise may thi-ther
fly; The earth is not too low, His prai-ses there may grow. Let
all the world in e-very cor-ner sing, My God and King!

2 Let all the world in every corner sing,
   My God and King!
   The Church with psalms must shout,
   No door can keep them out;
   But, above all, the heart
   Must bear the longest part.
   Let all the world in every corner sing,
   My God and King!

Descant for voices or recorders: John Whitworth, b.1921

Let all the world in e-very cor-ner sing, My God and King!
Let all the world in e - very cor-ner sing,
My God and King, my God and King! My God and King!

* *Sing or play either or both notes*

32

Prayers

God our Father, creator of the world, please help us to love one another. Make nations friendly with other nations; make all of us love one another like brothers. Help us to do our part to bring peace in the world and happiness to all men. *Amen*

Dear Father of the world family, please take care of all children everywhere. Keep them safe from danger and help them grow up strong and good. *Amen*

God, this is your world. You made us. Teach us how to live in the world that you have made. *Amen*

✠ 16 Now thank we all our God

M. Rinkart, 1586–1649,
translated by C. Winkworth, d.1878
Geoffrey Beaumont, 1903–70

GRACIAS

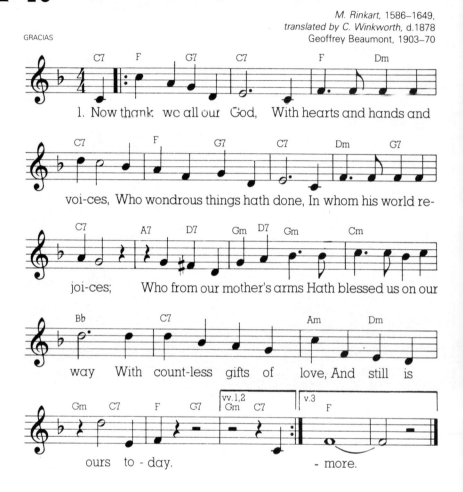

1. Now thank we all our God, With hearts and hands and voi-ces, Who wondrous things hath done, In whom his world re-joi-ces; Who from our mother's arms Hath blessed us on our way With count-less gifts of love, And still is ours to - day. - more.

2 O may this bounteous God
   Through all our life be near us,
   With ever joyful hearts
   And blessed peace to cheer us;
   And keep us in his grace,
   And guide us when perplexed,
   And free us from all ills
   In this world and the next.

3 All praise and thanks to God
   The Father now be given,
   The Son, and Him who reigns
   With Them in highest heaven;
   The One eternal God,
   Whom earth and heaven adore;
   For thus it was, is now,
   And shall be evermore.

34

Prayer

Father God we sing our praise and thanks to you:
    for you are our friend.
You love us and look after us,
    and nothing happens without your noticing.
You keep on being kind to us however little we deserve it.
In Jesus you show us the right way to behave:
    and if we trust you, you help us to live as your family.
So everyone who knows you thanks you and loves you.
Together we thank you now. *Amen*

✠ 17 Praise to the Lord

HAST DU DENN, JESU

J. Neander, 1650–80
Melody from *Stralsund Gesangbuch*, 1665

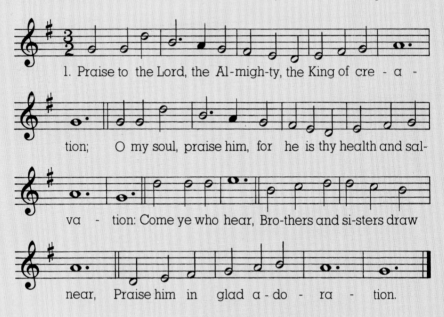

1. Praise to the Lord, the Al-migh-ty, the King of cre - a -
tion; O my soul, praise him, for he is thy health and sal-
va - tion: Come ye who hear, Bro-thers and si-sters draw
near, Praise him in glad a - do - ra - tion.

2 Praise to the Lord, who o'er all things so wondrously reigneth,
Shelters thee gently from harm or when fainting sustaineth:
    Hast thou not seen
    All thy heart's wishes have been
    Granted in what he ordaineth?

3 Praise to the Lord, who doth prosper thy work, and defend thee;
Surely his goodness and mercy doth daily attend thee:
    Ponder anew
    What the Almighty can do,
    He who with love doth befriend thee.

4 Praise to the Lord! O let all that is in me adore him!
All that hath life and breath come now with praises before him!
    Let the amen
    Sound from his people again:
    Gladly for ay we adore him.

Rejoice in the Lord always 18

Round in 2 parts

Anon

Re-joice in the Lord always, and a-gain I say re-joice,

Re-joice,   re-joice,   and a - gain I say re - joice.

Let us with a gladsome mind
praise the Lord for he is kind:
    for his mercies shall endure,
    ever faithful, ever sure.

All things living he does feed;
his full hand supplies their need:
    for his mercies shall endure,
    ever faithful, ever sure.

John Milton, 1608–74

✠ 19 O worship the King

✝ Sir R. Grant, 1785–1838
Supplement to the New Version, 1708
Probably by W. Croft, 1678–1727

HANOVER

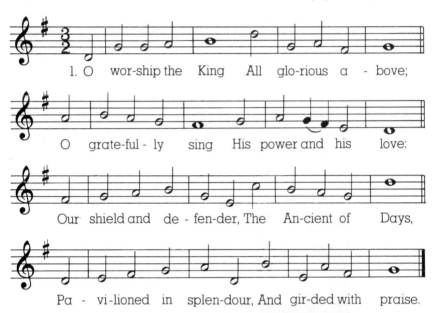

1. O wor-ship the King All glo-rious a - bove;

O grate-ful - ly sing His power and his love:

Our shield and de - fen-der, The An-cient of Days,

Pa - vi-lioned in splen-dour, And gir-ded with praise.

2 O tell of his might,
    O sing of his grace,
Whose robe is the light,
    Whose canopy space.
His chariots of wrath
    The deep thunder-clouds form,
And dark is his path
    On the wings of the storm.

3 This earth, with its store
    Of wonders untold,
Almighty, thy power
    Hath founded of old;
Hath stablished it fast
    By a changeless decree,
And round it hath cast,
    Like a mantle, the sea.

Praise and thanksgiving

20

Albert Frederick Bayley, b.1901
Gaelic Melody, arranged by
John Whitworth, b.1921

BUNESSAN

1. Praise and thanks-gi-ving, Fa-ther, we of - fer for all things

li-ving thou ma-dest good. Har-vest of sown fields, fruits of the

or-chard, hay from the mown fields, blos-som and wood.

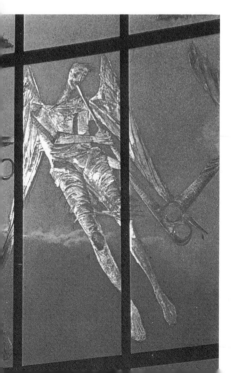

2 Father, providing
food for thy children,
thy wisdom guiding,
teach us to share
one with another,
so that rejoicing
with us, our brother
may know thy care.

3 Then will thy blessing
reach every people,
all men confessing
thy gracious hand.
Where thy will reigneth
no man will hunger,
thy love sustaineth;
fruitful the land.

✠ 21 Sing to the Lord

Marie Odile Herve
Richard H. Jacquet, b.1947

GOFF'S OAK

1. Sing to the Lord,
stars and beau - ti - ful,
sun, mil - lions of rain - drops, sing.
Brooks and ri-vers that run, sing to the Lord.

2 Sing to the Lord,
 rolling waves on the sand,
 seaweed and pebbles, sing.
 Mighty mountains that stand,
 sing to the Lord.

3 Sing to the Lord,
 wheat that sways in the breeze,
 ants ever busy, sing.
 Cheerful birds in the trees,
 sing to the Lord.

4 Sing to the Lord,
 playful kittens and lambs,
 mothers and children, sing.
 Smiling babies in prams,
 sing to the Lord.

5 Sing to the Lord,
 great and wonderful world,
 children and grown-ups, sing.
 Songs of praise for this world
 sing to the Lord.

The Prophet has said,
'All creatures are God's children,
and those dearest to God
are the ones who treat his children kindly.'

✠ 22 Sing a new song to the Lord

Timothy Dudley-Smith, b.1926
David G. Wilson, b.1940

PSALM 98

1. Sing a new song to the Lord, He to whom won-ders be-long. Re - joice in his tri-umph and tell of his power. O sing to the Lord a new song. song.

2 Now to the ends of the earth
See his salvation is shown.
And still he remembers his mercy and truth,
Unchanging in love to his own.

3 Sing a new song and rejoice.
Publish his praises abroad.
Let voices in chorus with trumpet and horn,
Resound for the joy of the Lord.

4 Join with the hills and the sea
Thunders of praise to prolong.
In judgement and justice he comes to the earth.
O sing to the Lord a new song.

Prayer

O sing a new song to the Lord: sing to the Lord all the earth.
O sing to the Lord, bless his name: proclaim his help day by day.
Give the Lord, you families of peoples,
Give the Lord glory and power:
Response: Give the Lord the glory of his name.

It is the Lord who has brought us together to pray, and praise him,
and receive his instruction.
Give the Lord glory and power:
Response: Give the Lord the glory of his name.

It is the Lord who has shown his great love for us,
and has taught us, through Christ, to call him 'Father'.
Give the Lord glory and power:
Response: Give the Lord the glory of his name. *Amen*

Based on the Gelineau translation of Psalm 96

23 The angel Gabriel

Sabine Baring-Gould, 1834–1924
Traditional,
arranged by John Whitworth, b.1921

1. The angel Gabriel from heaven came, His wings as drifted snow, his eyes as flame, 'All hail,' said he, 'thou lowly maiden Mary, Most highly favour'd lady.' Gloria!

2 'For known a blessed Mother thou shalt be,
All generations laud and honour thee,
Thy Son shall be Emmanuel, by seers foretold.
Most highly favour'd lady.' Gloria!

3 Then gentle Mary meekly bowed her head,
'To me be as it pleaseth God,' she said,
'My soul shall laud and magnify his Holy Name.'
Most highly favour'd lady. Gloria!

4 Of her, Emmanuel, the Christ, was born
In Bethlehem, all on a Christmas morn,
And Christian folk throughout the world will ever say –
Most highly favour'd lady. Gloria!

And in the sixth month the angel Gabriel was sent from God unto a City of Galilee, named Nazareth.

To a virgin espoused to a man whose name was Joseph, of the house of David, and the virgin's name was Mary.

And the angel said unto her, 'Fear not, Mary, for thou hast found favour with God. And behold thou shalt conceive in thy womb and bring forth a son and shalt call his name Jesus.'

Luke 1 vv 26, 27, 30

Mary's song of praise

My heart praises the Lord;
my soul is glad because of God my Saviour,
For he has remembered me, his lowly servant!
From now on all people will call me happy,
because of the great things the Mighty God has done for me.
His name is holy.

Luke 1 vv 46–49

We look forward to the coming of Jesus and prepare ourselves for it.
Advent means 'coming before'.

24 A is for Advent

Cyril G. Hambly
SWAFFHAM
Cyril G. Hambly, b. 1931

1. A is for Advent: Season of joy, Je - sus is coming:

Wor-ship this won-der-ful boy. might.

2 D is for Deliverer:
   Fighting the wrong,
   Jesus the Saviour;
   Join in a welcoming song.

3 V for his Visit:
   Banishing night,
   Jesus the Day-star,
   Candles of love are alight.

4 E for Excitement:
   Yes, he will come,
   Jesus expected,
   Open your hearts to God's son.

5 N for his Nearness:
   Comforting arm,
   Jesus our brother,
   Keeping his children from harm.

6 T for his Triumph:
   Triumph of right,
   Jesus is reigning,
   Trust in his conquering might.

A
IS FOR
ADVENT

COMING BEFORE

SEASON OF JOY

D FOR DELIVER

JESUS

V FOR
HIS VISIT

IS COMING

E FOR EXCITEMENT

WORSHIP THIS
WONDERFUL

N FOR
HIS NEARNESS

BOY

T FOR HIS TRIUMPH

47

25 Hark the glad sound!

P. Doddridge, 1702–51
BRISTOL
Melody from Ravenscroft's *Psalter*, 1621

1. Hark the glad sound! the Sa-viour comes, The Sa-viour pro-mised long; Let e-very heart pre-pare a throne, And e-very voice a song.

2 He comes the prisoners to release
   In Satan's bondage held;
   The gates of brass before him burst,
   The iron fetters yield.

3 He comes the broken heart to bind,
   The wounded soul to cure,
   And with the treasures of his grace
   To enrich the humble poor.

4 Our glad hosannas, Prince of Peace,
   Your welcome shall proclaim;
   And heaven's eternal arches ring
   With your beloved name.

Prayer

Lord God, we praise you for coming to us in the past.
You have taught us about yourself through the great men
   of the Bible.
You have shown us what you are like in the life of
   your son, Jesus.
Lord God, we praise you for coming to us now.
You come to us in the love of other people.
You come to us in the needs of other people.
You come to us as we worship you together.
We welcome you, the God who comes. *Amen*

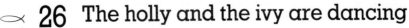

26 The holly and the ivy are dancing

Emily Chisholm, b.1910
Traditional, arranged by Alan D. Percival

1. The hol-ly and the i-vy Are dan-cing in a ring, Round the

ber-ry-bright red can-dles And the white and shi-ning King.

2 Oh, one is for the prophets
And for the light they bring.
They are candles in the darkness,
All alight for Christ the King.

3 And two for John the Baptist.
He calls on us to sing:
'O prepare the way of Jesus Christ,
He is coming, Christ the King.'

4 And three for Mother Mary.
'I cannot see the way,
But you promise me a baby.
I believe you. I obey.'

5 And four are for God's people
In every age and day.
We are watching for his coming.
We believe and we obey.

6 And Christ is in the centre,
For this is his birthday,
With the shining nights of Christmas
Singing, 'He has come today.'

Optional descant (vv. 2–6) for voices or recorders:

2. One is for the pro-phets And for the light they bring

All a - light for Christ the King.

A prophecy of Christ's coming

The people that walked in darkness have seen a great light: they that dwell in the land of the shadow of death, upon them hath the light shined.

Isaiah 9 v 2

Some people make an Advent ring which has four candles. They light one on the first Sunday of Advent, two on the second, and so on. The light of the candles reminds us of Jesus, who was called 'The Light of the World'.

Early Jewish prophets said that a king would come to look after his people. A later prophet, John the Baptist, prepared the way for Jesus to do his work. So we remember the story of John during Advent.

27 Long ago, prophets knew

Fred Pratt Green, b.1903
Melody from *Piae Cantiones*, 1582,
arranged by Gustav Holst, 1874–1934

THEODORIC

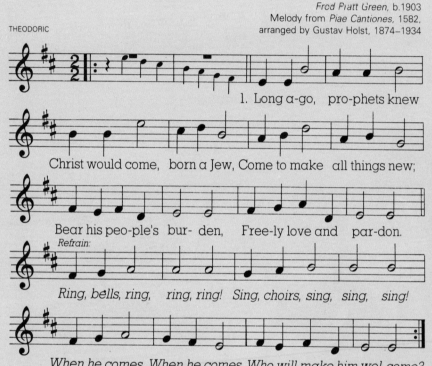

1. Long a-go, pro-phets knew
Christ would come, born a Jew, Come to make all things new;
Bear his peo-ple's bur- den, Free-ly love and par-don.

Refrain:

Ring, bells, ring, ring, ring! Sing, choirs, sing, sing, sing!

When he comes, When he comes, Who will make him wel-come?

2 God in time, God in man,
   This is God's timeless plan:
   He will come, as a man,
   Born himself of woman,
   God divinely human.
   Refrain:

3 Mary hail! Though afraid.
   She believed, she obeyed.
   In her womb, God is laid;
   Till the time expected,
   Nurtured and protected.
   Refrain:

4 Journey ends! Where afar
   Bethl'em shines, like a star,
   Stable door stands ajar.
   Unborn Son of Mary,
   Saviour, do not tarry!
   Refrain:

Prayer

O God, you have sent your servants to prepare your way. Fill our hearts with love, and strengthen our hands to work, that we may make ready the way for the coming of Jesus at Christmas. *Amen*

28 Every star shall sing a carol

Sydney Carter
Sydney Carter, b.1915

1. E-very star shall sing a ca-rol,
E - very crea-ture high or low, Come and praise the
King of Hea-ven By what-e - ver name you know.

Refrain:
God a-bove, Man be-low, Ho-ly is the name I know.

2 When the King of all creation
  Had a cradle on the earth,
  Holy was the human body,
  Holy was the human birth.
  Refrain:

3 Every star and every planet,
  Every creature high and low,
  Come and praise the King of Heaven
  By whatever name you know.
  Refrain:

We celebrate the birth of Jesus each year on 25 December.

And she brought forth her first born son and wrapped him in swaddling clothes and laid him in a manger; because there was no room for them in the inn.

Luke 2 v 7

29 Away in a manger

CRADLE SONG

Anon
W. J. Kirkpatrick, 1838–1921

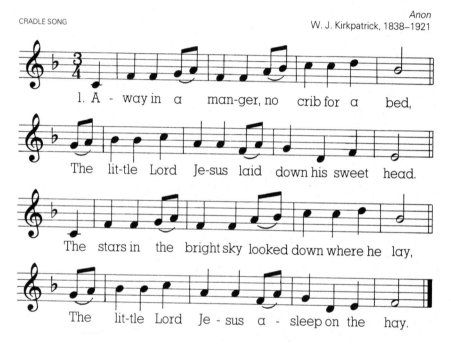

1. A - way in a man-ger, no crib for a bed,
The lit-tle Lord Je-sus laid down his sweet head.
The stars in the bright sky looked down where he lay,
The lit-tle Lord Je - sus a - sleep on the hay.

2 The cattle are lowing, the baby awakes,
   But little Lord Jesus no crying he makes.
   I love thee, Lord Jesus! look down from the sky,
   And stay by my side until morning is nigh.

3 Be near me, Lord Jesus; I ask thee to stay
   Close by me for ever, and love me, I pray.
   Bless all the dear children in thy tender care,
   And fit us for heaven to live with thee there.

Keeping Christmas

How will you your Christmas keep?
Feasting, fasting or asleep?
Will you laugh or will you pray,
Or will you forget the day?

Be it kept with joy or prayer,
Keep of either some to spare;
Whatsoever brings the day,
Do not keep but give away.

Eleanor Farjeon

30 Go tell it on the mountain

Negro Spiritual,
arranged by H. T. Burleigh, 1866–1949

2 And lo, when they had seen it, they all bowed down and prayed,
   they travelled on together to where the Babe was laid . . .
   Refrain:

3 When I was a seeker, I sought both night and day:
   I asked my Lord to help me and he showed me the way . . .
   Refrain:

4 He made me a watchman upon the city wall,
   And if I am a Christian, I am the least of all . . .
   Refrain:

Infant holy 31

Translated by Edith M. G. Reed, 1885–1933
Traditional Polish,
arranged by John Whitworth, b. 1921

1. In-fant ho-ly, in-fant low-ly, For his bed a cat-tle stall;
O-xen low-ing lit-tle know-ing Christ the babe is Lord of all.
Swift are winging angels singing, Nowells ringing tidings bringing,
Christ the babe is Lord of all, Christ the babe is Lord of all.

2 Flocks were sleeping, shepherds keeping
Vigil till the morning new,
Saw the glory, heard the story,
Tidings of a gospel true.
Thus rejoicing, free from sorrow,
Praises voicing, greet the morrow,
Christ the babe was born for you,
Christ the babe was born for you.

Prayer

O Lord Jesus Christ, bless all who try to bring happiness
to the hearts of others during this Christmastime; let
your loving spirit be felt throughout the world; and help
us to remember that it is better to give things to others
than only to receive things ourselves. We ask this for the
sake of Jesus Christ. *Amen*

32 Now tell us, gentle Mary

Translated from the French
by W. B. Lindsay
Arranged by Ruth Heller

Allegretto ♩ = 126

1. Now tell us, gentle Mary, what did Gabriel say to you? Now tell us of the ti-dings that he brought to Ga-li - lee. He told me I was fa-vour'd, That I would be the one _ God chose to be the mo-ther of Je - sus, his own son.

2 Now tell us, gentle Mary,
Of the birth of Christ that morn.
Now tell us of Christ Jesus,
Where it was that he was born.
Not in a palace glorious,
Not in a silken bed,
But in a humble stable
Did Jesus lay his head.

Now the holly bears a berry 33

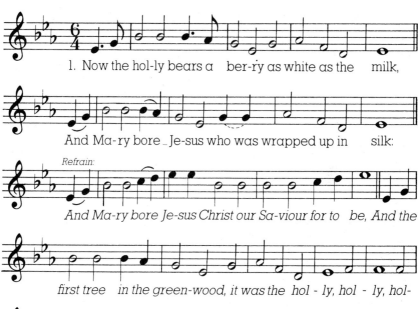

SANS DAY

Percy Dearmer, 1867–1936
Traditional English,
arranged by Martin Shaw, 1875–1958

1. Now the hol-ly bears a ber-ry as white as the milk,

And Ma-ry bore _ Je-sus who was wrapped up in silk:

Refrain:
And Ma-ry bore Je-sus Christ our Sa-viour for to be, And the

first tree in the green-wood, it was the hol - ly, hol - ly, hol-

-ly. And the first tree in the green-wood, it was the hol - ly.

2 Now the holly bears a berry as green as the grass,
   And Mary bore Jesus, who died on the cross:
   Refrain:

3 Now the holly bears a berry as black as the coal,
   And Mary bore Jesus, who died for us all:
   Refrain:

4 Now the holly bears a berry, as blood is it red,
   Then trust we our Saviour, who rose from the dead:
   Refrain:

61

34 O little town of Bethlehem

FOREST GREEN

Bishop Philip Brooks, 1835–93
Traditional English, arranged,
collected, and adapted by Ralph Vaughan Williams, 1872–1958

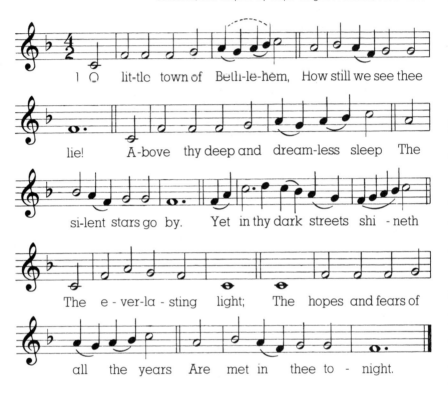

1 O lit-tle town of Beth-le-hem, How still we see thee lie! A-bove thy deep and dream-less sleep The si-lent stars go by. Yet in thy dark streets shi - neth The e - ver-la - sting light; The hopes and fears of all the years Are met in thee to - night.

2 O morning stars, together
    Proclaim the holy birth,
And praises sing to God the King,
    And peace to men on earth;
For Christ is born of Mary;
    And, gathered all above,
While mortals sleep, the angels keep
    Their watch of wondering love.

3 How silently, how silently,
   The wondrous gift is given!
So God imparts to human hearts
   The blessings of his heaven.
No ear may hear his coming;
   But in this world of sin,
Where meek souls will receive him, still
   The dear Christ enters in.

4 O holy Child of Bethlehem,
   Descend to us, we pray;
Cast out our sin, and enter in,
   Be born in us to-day.
We hear the Christmas Angels
   The great glad tidings tell:
O come to us, abide with us,
   Our Lord Emmanuel.

Christmas poem

Soft as silent snow Lord Jesus came,
Soft as Mary's voice telling his name,
Softly chimed a bell in Bethlehem.

Soft as the muzzle of the little ass,
Soft, the new white lamb upon the grass
And softly chimed a bell that first Christmas.

Jennifer Andrews

Prayer

God our Father, we thank you for all the joys of Christmas
but we remember, at this happy time, all those who have
no homes, those who are hungry, and all those people
who have no family and friends. Please take care of them
and help us to do all we can to make their lives happier. *Amen*

35 O come, all ye faithful

† *18th Century, translated by F. Oakeley, W. T. Brooke, and others*

ADESTE FIDELES

Anon

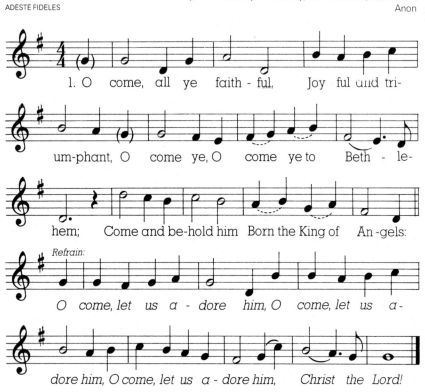

1. O come, all ye faith - ful, Joy ful and tri - um-phant, O come ye, O come ye to Beth - le - hem; Come and be-hold him Born the King of An - gels:

Refrain:

O come, let us a - dore him, O come, let us a - dore him, O come, let us a - dore him, Christ the Lord!

2 See how the Shepherds,
Summoned to his cradle,
Leaving their flocks, draw nigh with lowly fear;
We too will thither
Bend our joyful footsteps:
Refrain:

3 Sing, choirs of Angels.
Sing in exultation,
Sing, all ye citizens of heaven above;
Glory to God
In the Highest:
Refrain:

4 Yea, Lord, we greet thee,
Born this happy morning,
Jesu, to thee be glory given;
Word of the Father,
Now in flesh appearing:
Refrain:

Prayer

O God, our Father, we join the angels to praise you for
your great and wonderful gift to the world. As we
remember the birth of Jesus, fill our hearts with gladness
and the spirit of goodwill, so that we may praise you both
in what we say and the things we do. *Amen*

36 Once in royal David's city

IRBY

† Mrs C. F. Alexander, 1818–95
H. J. Gauntlett, 1805–1876,
arranged by A. H. Mann, 1850–1930

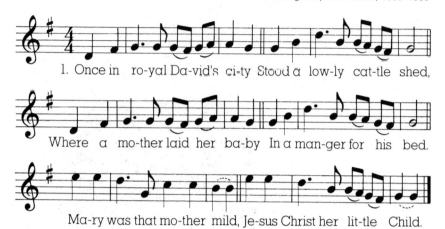

1. Once in ro-yal Da-vid's ci-ty Stood a low-ly cat-tle shed,

Where a mo-ther laid her ba-by In a man-ger for his bed.

Ma-ry was that mo-ther mild, Je-sus Christ her lit-tle Child.

2 He came down to earth from heaven,
    Who is God and Lord of all,
And his shelter was a stable,
    And his cradle was a stall;
With the poor, and mean, and lowly,
Lived on earth our Saviour holy.

3 For he is our childhood's pattern,
    Day by day like us he grew,
He was little, weak, and helpless,
    Tears and smiles like us he knew;
And he feeleth for our sadness,
And he shareth in our gladness.

4 And our eyes at last shall see him,
    Through his own redeeming love,
For that child so dear and gentle
    Is our Lord in heaven above;
And he leads his children on
To the place where he is gone.

See him lying on a bed of straw **37**

Michael Perry, b.1942

CALYPSO CAROL

Michael Perry, arranged by Stephen Coates

1. See him ly-ing on a bed of straw; A draugh-ty sta-ble with an o-pen door; Ma-ry cra-dl-ing the Babe she bore; The Prince of Glo-ry is his name. *O now carry me to Beth-le-hem To see the Lord ap-pear to men: Just as poor as was the sta-ble then, The Prince of Glo-ry when he came.*

Ba-by can be sal-va-tion to the soul.

Star of silver sweep across the skies,
Show where Jesus in the manger lies,
Shepherds swiftly from your stupor rise
To see the Saviour of the world.
Refrain:

3 Angels, sing again the song you sang,
Bring God's glory to the heart of man:
Sing that Bethlehem's little Baby can
Be salvation to the soul.
Refrain:

38 The Virgin Mary had a baby boy

Melody from the Edric Connor Collection,
arranged by D. J. Crawshaw

1. The Vir-gin Ma-ry had a ba-by boy. The
Vir-gin Ma-ry had a ba-by boy, The Vir-gin Ma-ry had a
ba-by boy, And they say that his name was Je-sus.

Refrain:

He come from the glo-ry, He come from the
glo-ri-ous king-dom; O yes! be-lie-ver, O yes! be-lie-ver.

He come from the glory, He come from the glorious kingdom.

2 The angels sang when the baby was born,
The angels sang when the baby was born,
The angels sang when the baby was born,
And proclaimed him the Saviour Jesus.
Refrain:

3 The wise men saw where the baby was born,
The wise men saw where the baby was born,
The wise men saw where the baby was born,
And they saw that his name was Jesus.
Refrain:

Under Bethlem's star **39**

M. C. Vojáček
Traditional Czech, arranged by Leslie Russell, b.1901

1. Un - der Beth - lem's star so bright,
Shepherds watched their flocks by night. *Hy-dom, hy-dom,*

Refrain:

vv. 1,2,3,4 | v.5

hydomdom, Hydom, hydom, hydomdom. hydomdom (hum)

2 Came an angel telling them,
They must go to Bethlehem.
Refrain:

3 'Hasten, hasten,' they did say,
'Jesus Christ you'll find that way.'
Refrain:

4 'Sleeping in a manger bare
Lies the Holy Child so fair.'
Refrain:

5 'Mary rocks him tenderly.
Joseph sings a lullaby.'
Refrain:

Epiphany, which is on 6 January, is the time when we celebrate the visit of the wise men to the house in Bethlehem where the family of Jesus stayed after his birth. They brought gifts fit for a king.

40 Lift up lightly the stable bar

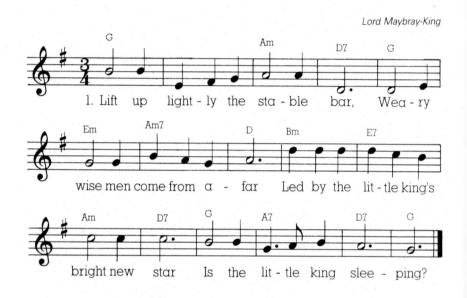

Lord Maybray-King

1. Lift up light-ly the sta-ble bar, Wea-ry wise men come from a - far Led by the lit-tle king's bright new star Is the lit-tle king slee - ping?

2. Golden gifts for a baby king,
   Cloth of gold and a golden ring,
   Set by the cradle this gift I bring
   Is the little king sleeping?

3. Frankincense for a priest divine,
   Born on this first Christmas time,
   Set by the cradle this gift of mine
   Is the little king sleeping?

4. Myrrh for bitterness yet to be,
   Let it not trouble his infancy.
   Hush, my companions, and pray with me
   Is the little king sleeping?

Descant recorder

Now when Jesus was born in Bethlehem of Judaea in the days of Herod the king, behold, there came wise men from the east to Jerusalem: saying, 'Where is he that is born king of the Jews for we have seen his star in the east and are come to worship him.'

Matthew 2.1

Prayer

Lord God, as the wise men brought their gifts to Jesus, so may we offer our time, the things we have and all the things we can do, in your service and for the benefit of others. *Amen*

71

41 Unto us a boy is born

15th century carol, translated by Percy Dearmer, 1867–1936
From Piae Cantiones, *1582,*
arranged by Geoffrey Shaw, 1879–1943

PUER NOBIS NASCITUR

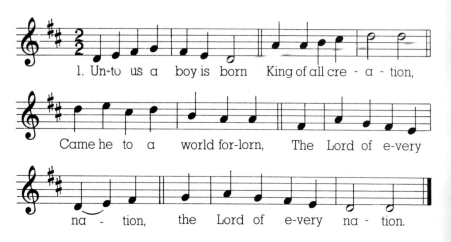

1. Un-to us a boy is born King of all cre - a - tion,

Came he to a world for-lorn, The Lord of e-very

na - tion, the Lord of e-very na - tion.

2 Cradled in a stall was he
   With sleepy cows and asses;
  But the very beasts could see
   That he all men surpasses.
   That he all men surpasses.

3 Herod then with fear was filled:
   'A prince', he said, 'in Jewry!'
  All the little boys he killed
   At Bethlem in his fury.
   At Bethlem in his fury.

4 Now may Mary's son, who came
   So long ago to love us,
  Lead us all with hearts aflame
   Unto the joys above us.
   Unto the joys above us.

Lent begins on Ash Wednesday, which is the day after Shrove Tuesday (Pancake Day). It is the time when Jesus spent forty days alone in a rocky wilderness thinking about how he was to do God's work.

42 Forty days and forty nights

✝ G. H. Smyttan, 1825–70, and F. Pott
AUS DER TIEFE
Martin Herbst, 1654–81

1. For - ty days and for - ty nights
Thou wast fa - sting in the wild; For -ty days and
for - ty nights Temp-ted, and yet un - de - filed:

2 Sunbeams scorching all the day;
      Chilly dew-drops nightly shed;
   Prowling beasts about thy way;
      Stones thy pillow, earth thy bed.

3 Keep, O keep us, Saviour dear,
      Ever constant by thy side;
   That with thee we may appear
      At the eternal Eastertide.

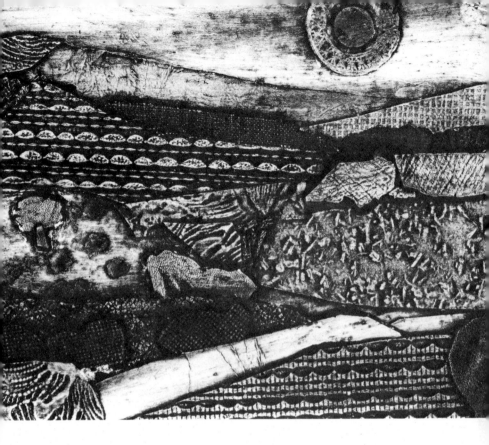

Prayer

Most wonderful and loving God, forgive us for everything that is
wrong in our lives.
By the power of the Holy Spirit help us to live as Jesus lived,
so that we may say thank you for your gifts in everything we say and
do.
We ask it in his name. *Amen*

This is how God's love has been made clear to us: he sent his only son
to live among us to help us to live splendidly. He loved us enough to
send his son to help us to get rid of all that is wrong in our hearts and
lives. That's how we know what love means.

1 John 4 vv 9–10

43 All glory, laud and honour

Theodulph of Orleans, d.821, translated by J. M. Neale
Melody by M. Teschner, 1613,
arranged by J. S. Bach, 1688–1750

ST THEODULPH

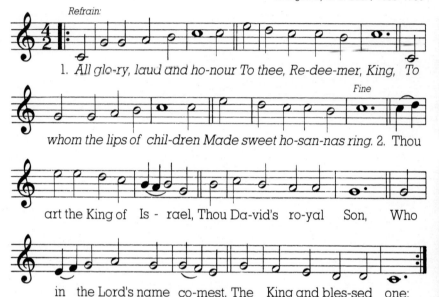

Refrain:

1. All glo-ry, laud and ho-nour To thee, Re-dee-mer, King, To

Fine

whom the lips of chil-dren Made sweet ho-san-nas ring. 2. Thou

art the King of Is - rael, Thou Da-vid's ro-yal Son, Who

in the Lord's name co-mest, The King and bles-sed one:

3 The company of angels
        Are praising thee on high,
    And mortal men and all things
        Created make reply: *Refrain:*

4 The people of the Hebrews
        With palms before thee went;
    Our praise and prayer and anthems
        Before thee we present: *Refrain:*

5 To thee before thy passion
        They sang their hymns of praise;
    To thee now high exalted
        Our melody we raise: *Refrain:*

6 Thou didst accept their praises:
        Accept the prayers we bring,
    Who in all good delightest,
        Thou good and gracious King: *Refrain:*

Palm Sunday is the beginning of Holy Week which was the last week of Jesus' life on earth.

Crowds of people carpeted the road with their cloaks, and some cut branches from the trees to spread in his path. Then the crowd that went ahead and the others that came behind raised the shout: 'Hosanna to the Son of David! Blessings on him who comes in the name of the Lord!'

Matthew 21 vv 5, 8, 9

'Here is your king, who comes to you in gentleness, riding on an ass, riding on the foal of a beast of burden!'

Jesus rode into Jerusalem on a humble donkey instead of on a fine horse such as a king would be expected to ride.

44 Ride on! ride on in majesty!

† H. H. Milman, 1791–1868
Adapted from Chorale in
Musikalisches Handbuch, Hamburg, 1690

WINCHESTER NEW

1. Ride on! ride on in ma - je - sty! Hark, all the tribes ho - san-na cry; O Sav-iour meek pur - sue thy road With palms and scat-tered gar- ments strowed.

2 Ride on! ride on in majesty!
   The angel armies of the skies
   Look down with sad and wondering eyes
   To see the approaching sacrifice.

3 Ride on! ride on in majesty!
   Thy last and fiercest strife is nigh;
   The Father, on his sapphire throne,
   Expects his own anointed Son.

4 Ride on! ride on in majesty!
   In lowly pomp ride on to die;
   Bow thy meek head to mortal pain,
   Then take, O God, thy power, and reign.

Prayer of the donkey

O God who made me
to trudge along the road
always,
to carry heavy loads
and to be beaten
always!
Give me great courage and gentleness.
One day let someone understand me
that I may no longer want to weep
because I never say what I mean
and they make fun of me.
Let me find a juicy thistle –
and make them give me time to pick it.
And, Lord, one day let me find again
my little brother of the Christmas crib.

Rumer Godden

45 Trotting, trotting through Jerusalem

Eric Reid
Eric Reid, 1936–1970

1 Trotting, trotting through Je-rusalem, Jesus, sitting on a don-key's back, Chil-dren wa-ving bran-ches, sing-ing, 'Happy is he that comes in the name of the Lord!' Lord!'

2 Many people in Jerusalem
   Thought he should have come on a mighty horse
   Leading all the Jews to battle –
   'Happy is he that comes in the name of the Lord!'

3 Many people in Jerusalem
   Were amazed to see such a quiet man
   Trotting, trotting on a donkey
   'Happy is he that comes in the name of the Lord!'

4 Trotting, trotting through Jerusalem,
   Jesus, sitting on a donkey's back,
   Let us join the children singing
   'Happy is he that comes in the name of the Lord!'

Prayer

Lord, we remember how people cheered when Jesus rode into Jerusalem. Help us also to praise him joyfully and to go on praising him even when life is difficult and things go wrong. *Amen*

The donkey

When fishes flew and forests walked
And figs grew upon thorn,
Some moment when the moon was blood
Then surely I was born;

With monstrous head and sickening cry
And ears like errant wings,
The devil's walking parody
On all four footed things.

The tattered outlaw of the earth,
Of ancient crooked will;
Starve, scourge, deride me; I am dumb,
I keep my secret still.

Fools! For I also had my hour;
One far fierce hour and sweet:
There was a shout about my ears,
And palms before my feet.

G. K. Chesterton

Jesus was crucified on the Friday of Holy Week. We call this Friday 'Good' because Jesus showed his love for men by being prepared to die for us all.

46 There is a green hill

HORSLEY

Mrs C. F. Alexander, 1818–95
William Horsley, 1774–1858

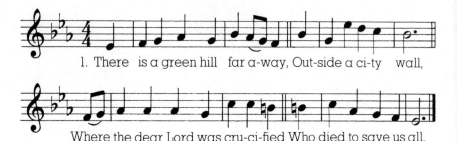

1. There is a green hill far a-way, Out-side a ci-ty wall,

Where the dear Lord was cru-ci-fied Who died to save us all.

2 We may not know, we cannot tell,
    What pains he had to bear,
  But we believe it was for us
    He hung and suffered there.

3 He died that we might be forgiven,
    He died to make us good;
  That we might go at last to heaven,
    Saved by his precious blood.

4 O, dearly, dearly has he loved,
    And we must love him too,
  And trust in his redeeming blood,
    And try his works to do.

And when they came to the place which is called The Skull, there they crucified him, and the criminals, one on the right, and one on the left. And Jesus said, 'Father, forgive them; for they know not what they do.'

Luke 23 vv 33–34

This is my commandment, that you love one another as I have loved you. Greater love has no man than this, that a man lay down his life for his friends.

John 15 vv 12–13

47 Who was the other who died?

GOOD FRIDAY

✝ Cecily Taylor
Peter D. Smith, b.1938 and Cecily Taylor, b.1930

1. Who was the o-ther who died on the hill? Who was the o-ther closed in for the kill? One was a rob-ber and one was a thief But who was the third man whose life was so brief? *Who do men say that I am? Who do men say that I am? Who do you say that I am?*

2 Some say a prophet come back from the dead,
Some an idealist but rather misled;
Some say a teacher or King of the Jews,
And some say God's son who had no power to choose.
Who do men say that I am?
Who do men say that I am?

3 Who was the other who died on the hill?
Who was the other closed in for the kill?
One was a robber and one was a thief –
But who was the third man whose life was so brief?
Who do you say that I am?
Who do you say that I am?

When Jesus was crucified the soldiers nailed a notice above his head which said, 'Jesus of Nazareth, King of the Jews'. This was written in three languages. The Latin words were 'Iesus Nazarenus Rex Judaeorum'. You can sometimes see the first letters of these words over a crucifix.

Easter is the festival when we celebrate Jesus returning from death to share a new life with his friends.

48 Jesus Christ is risen today

EASTER HYMN

Lyra Davidica, 1708, *and the Supplement*, 1816
Altered from melody in *Lyra Davidica*, 1708

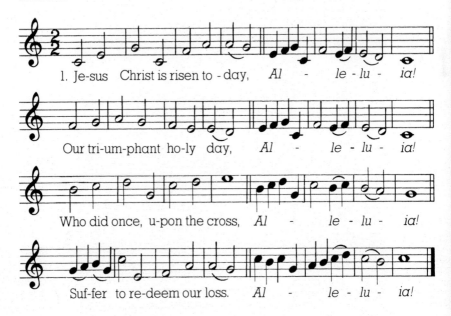

1. Je-sus Christ is risen to-day, *Al - le - lu - ia!*
Our tri-um-phant ho-ly day, *Al - le - lu - ia!*
Who did once, u-pon the cross, *Al - le - lu - ia!*
Suf-fer to re-deem our loss. *Al - le - lu - ia!*

2 Hymns of praise then let us sing, *Alleluia!*
Unto Christ, our heavenly King, *Alleluia!*
Who endured the cross and grave, *Alleluia!*
Sinners to redeem and save: *Alleluia!*

3 But the pains that he endured, *Alleluia!*
Our salvation have procured; *Alleluia!*
Now above the sky he's King, *Alleluia!*
Where the angels ever sing: *Alleluia!*

Prayer

Good Friday is a time of sadness,
Easter is a time of gladness,
On Good Friday Jesus died,
But rose again at Eastertide.
All thanks and praise to God. *Amen*

This is what we proclaim, that Christ was raised from the dead.

I Corinthians 15v12

87

The word Easter comes from the name of a Saxon goddess of the spring, Oeastre. Jesus died and rose from the dead during our springtime, bringing new life to his friends just as the spring goddess was said to bring new life to the dead world of winter.

49 Good Christian men

C. A. Alington, 1872–1955
Melody by Melchior Vulpius, c.1560–1615

VULPIUS

1. Good Chris-tian men, re - joice and sing.
Now is the tri - umph of our King. To all the
world glad news we bring: Al - le - lu - ia,
al - le - lu - ia, al - le - lu - ia!

2 The Lord of life has won the day,
Bring flowers of song to strew his way;
Let all mankind rejoice and say:
Alleluia!

3 Praise we in songs of victory
That love, that life which cannot die,
And sing with hearts uplifted high:
Alleluia!

4 Your name we bless, O risen Lord,
And sing today with one accord
The life laid down, the life restored:
Alleluia!

It is the joyful Easter time 50

A. M. Milner-Barry, 1875–1940
Traditional English

CORNISH CAROL

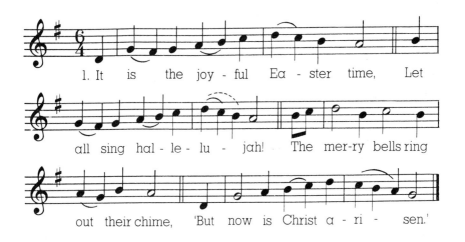

1. It is the joy-ful Ea-ster time, Let all sing hal-le-lu-jah! The mer-ry bells ring out their chime, 'But now is Christ a-ri-sen.'

2 The world is bright with flowers gay,
   And all Christ's people praise and pray,
   For Jesus rose on Easter Day;
   Sing joyful hallelujah!

51 Now the green blade riseth

NOEL NOUVELET

J. M. C. Crum, 1872–1958
Traditional French, arranged by
John Whitworth, b.1921

1 Now the green blade ri - seth from the bu-ried grain,

Wheat that in the dark earth ma-ny days has lain;

Love lives a - gain, that with the dead has been:

Love is come a - gain like wheat that spring-eth green.

2 In the grave they laid him, Love whom men had slain,
thinking that never he would wake again,
laid in the earth like grain that sleeps unseen:
love is come again like wheat that springeth green.

3 Forth he came at Easter, like the risen grain,
he that for three days in the grave had lain,
quick from the dead my risen Lord is seen:
love is come again like wheat that springeth green.

4 When our hearts are wintry, grieving or in pain,
thy touch can call us back to life again,
fields of our heart that dead and bare have been:
love is come again like wheat that springeth green.

The Golden Boy

In March he was buried
    And nobody cried
Buried in the dirt
    Nobody protested
Where grubs and insects
    That nobody knows
With outer-space faces
    That nobody loves
Can make him their feast
    As if nobody cared.

But the Lord's mother
    Full of her love
Found him underground
    And wrapped him with love
As if he were her baby
    Her own born love
She nursed him with miracles
    And starry love
And he began to live
    And to thrive on her love

He grew night and day
    And his murderers were glad
He grew like a fire
    And his murderers were happy
He grew lithe and tall
    And his murderers were joyful
He toiled in the fields
    And his murderers cared for him
He grew a gold beard
    And his murderers laughed.

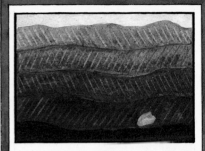

With terrible steel
    They slew him in the furrow
With terrible steel
    They beat his bones from him
With terrible steel
    They ground him to powder
They baked him in ovens
    They sliced him on tables
They ate him they ate him
    They ate him they ate him

Thanking the Lord
Thanking the Wheat
Thanking the Bread
For bringing them Life
Today and Tomorrow
Out of the dirt.

Ted Hughes

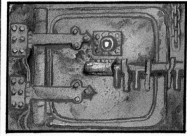
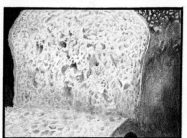

52 When Easter to the dark world came

W. H. Hamilton, 1886–1958
John Whitworth, b.1921

1. When Ea-ster to the dark world came, Fair
flowers glowed like scar-let flame: *At Ea-ster-tide, at
Ea-ster-tide, O glad was the world at Ea-ster-tide.*

2 When Mary in the garden walked,
And with her risen Master talked:
Refrain:

3 When John and Peter in their gloom
Met angels at the empty tomb:
Refrain:

4 When Thomas' heart with grief was black,
Then Jesus like a King came back:
Refrain:

5 And friend to friend in wonder said:
'The Lord is risen from the dead!'
Refrain:

6 This Eastertide with joyful voice
We'll sing, 'The Lord is King! Rejoice!'
Refrain:

It is true: the Lord has risen; he has appeared to Simon.

Luke 24 v 34

The Pascal candle

Some churches light a large candle at midnight on Easter Saturday night as a reminder of Jesus who was called 'the Light of the World'. It remains lit for forty days between Easter and Ascension Day.

Ascension Day is always held on a Thursday. Between Easter and Ascension Jesus appeared many times to his friends to show them that he would always be with them. On this day he showed them that he was also going to be with God the Father for ever as King of the Universe.

53 Rejoice, the Lord is King

Charles Wesley, 1707–88
GOPSAL
G. F. Handel, 1685–1759

1. Re-joice, the Lord is King, Your Lord and King a-dore; Mor-
tals, give thanks and sing, And tri-umph e - ver-more: *Lift*
up your heart, lift up your voice; Re-joice, a-gain I say, re-joice.

2 Jesus, the Saviour, reigns,
   The God of truth and love;
   When he had purged our stains,
   He took his seat above:
   Refrain:

3 His kingdom cannot fail;
   He rules o'er earth and heaven;
   The keys of death and hell
   Are to our Jesus given:
   Refrain:

4 He sits at God's right hand
   'Till all his foes submit,
   And bow to his command,
   And fall beneath his feet:
   Refrain:

I am with you always, to the end of time.

Matthew 28 v 20

And when he had spoken, while they beheld, he was taken up; and a cloud received him out of their sight.

Acts 1 v 9

Prayer

Jesus, when I am afraid, help me to remember that you are with me, nearer than my breathing, closer than my beating heart. Let me trust in you and help me to help others in their fears as you support me. *Amen*

The Holy Spirit is what we call God when he is active in a powerful and loving way in the world.

54 Come to us, creative Spirit

ANGEL VOICES

David Mowbray, b.1938
E. G. Monk, 1819–1900

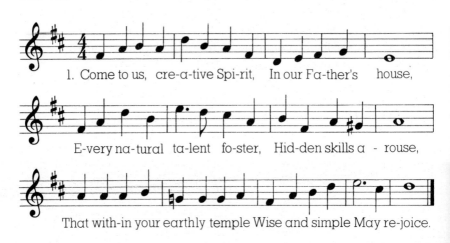

1. Come to us, cre-a-tive Spi-rit, In our Fa-ther's house,

E-very na-tural ta-lent fo-ster, Hid-den skills a - rouse,

That with-in your earthly temple Wise and simple May re-joice.

2 Poet, painter, music-maker,
All your treasures bring;
Craftsman, actor, graceful dancer,
Make your offering:
Join your hands in celebration!
Let Creation
Shout and sing!

3 Word from God Eternal springing
Fill our minds, we pray,
And in all artistic vision
Give integrity.
May the flame within us burning
Kindle yearning
Day by day.

4 In all places and forever
Glory be expressed
To the Son, with God the Father,
And the Spirit blest.
In our worship and our living
Keep us striving
Towards the best.

A thanksgiving

For all skilled hands, both delicate and strong –
Doctors' and nurses', soothing in their touch:
Sensitive artist-hands: musicians' hands
Vibrant with beauty: for all the hands that guide
Great ships amid great seas: for all brave hands,
Where'er they be, that ply their busy trades
With daily courage: homely mother-hands
Busy with countless different tasks each day:
For miners' hands that labour for our sakes:
For all hands rough and hard with honest work,
For old hands, frail and lovely, interlaced
With tell-tale wrinkles left by work and age:
For these we thank thee, Lord. *Amen*

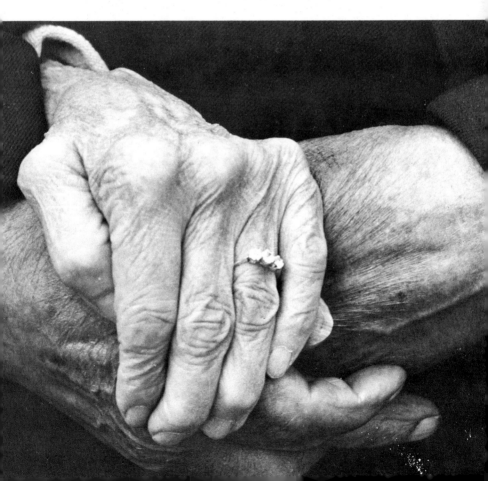

We cannot see God but we can see his Holy Spirit at work in the world. This is why the Holy Spirit is sometimes compared with the wind which is invisible but of great power. The Holy Spirit is also pictured as being like fire. Fire gives warmth and light.

55 Fire is lighting

John B. Geyer, b.1932
Allen Percival, b.1925

LIFELIGHT

In march time

Fire is ligh-ting torch and lamp at night; Fire out-

-bursts in-to power and light. Come, O God, Cre-a-tor, Spi-rit

now Fill all our lives with your fire.

2 Wind is battering waves of sea on land;
Wind is grinding the rocks to sand.
Come, O God, Creator, Spirit, now
Order anew all your world.

3 Water gushes down the cleft of space;
Living water and spring of grace.
Come, O God, Creator, Spirit, now
Grant us your life and your light.

And suddenly there came a sound from heaven as of a rushing mighty wind, and it filled all the house where the disciples were sitting.

And there appeared unto them cloven tongues like as of fire, and it sat upon each of them, and they were all filled with the Holy Spirit.

Acts 2 vv 2–3

Prayer

We thank you, Heavenly Father,
that when Jesus went back to be with you in heaven
you sent us the Holy Spirit to take his place.
Though we cannot see him, we know he is at work
in the world in everything that is good and holy,
and in our lives to carry out your will.
Send us the Holy Spirit, we pray,
to shape and mould our lives
and guide us day by day. *Amen*

The Dove is another symbol of the Holy Spirit as it looks and sounds like a very gentle bird and is white in colour. The Holy Spirit is not only powerful like the wind but is also gentle, pure and good.

56 There's a spirit in the air

+ Brian Wren, b.1936
Mediaeval French, arranged by
Ralph Vaughan Williams, 1872–1958

ORIENTIS PARTIBUS

1. There's a spi - rit in the air, tel-ling Chris-tians ev - ery - where, 'Praise the love that Christ re-vealed, li - ving, wor - king, in our world'.

2 Lose your shyness, find your tongue,
  tell the world what God has done:
    God in Christ has come to stay.
    We can see his power today.

3 When believers break the bread,
  when a hungry child is fed,
    praise the love that Christ revealed,
    living, working, in our world.

4 When a stranger's not alone,
  where the homeless find a home,
    praise the love that Christ revealed,
    living, working, in our world.

5 May his Spirit fill our praise,
  guide our thoughts and change our wo
    God in Christ has come to stay.
    We can see his power today.

6 There's a Spirit in the air,
  calling people everywhere:
    praise the love that Christ revealed,
    living, working, in our world.

Prayer

Father, as we go to our homes and our work this week
we ask you to send the Holy Spirit into our lives.

Open our ears – to hear what you are saying to us in the things
that happen to us and in the people we meet.
Open our eyes – to see the needs of the people around us.
Open our hands – to do our work well and to help when help is needed
Open our lips – to tell others the good news about Jesus and
bring comfort, happiness and laughter to other people.
Open our minds – to discover new truth about you and the world.
Open our hearts – to love you and our fellow men as you have
loved us in Jesus. *Amen*

It was on the first Whitsunday, which was on the Jewish Feast of Pentecost, that the followers of Jesus were given the courage to go out and tell people about him. Many believed them and were baptized and became members of the Church. Whitsunday is therefore the birthday of the Church.

57 Upon the day of Pentecost

Patricia Hunt, b.1921
Allen Percival, b.1925

GUIDANCE

1. U-pon the day of Pen-te-cost The
Ho-ly Spi-rit came Like power-ful, ru-shing,
migh-ty wind And lea-ping, li-ving flame.

2 The friends of Jesus till that hour
  Were fearful folk and weak;
  But now the Holy Spirit made
  Them bold and wise to speak.

3 With joy and confidence they went
  To all whom they could reach,
  In God the Holy Spirit's power
  To praise and heal and teach.

4 God's Holy Spirit still is here
  To guide our world today,
  And helps the friends of Jesus Christ
  In what they do and say.

Optional voices or instruments:

Up - on the day of Pen-te-cost The
Ho - ly Spi-rit came Like power-ful, rush - ing,
migh - ty wind And leap-ing, liv - ing flame.

58 I danced in the morning

Sydney Carter
Adapted from a Shaker melody
by Sydney Carter, b.1915

LORD OF THE DANCE

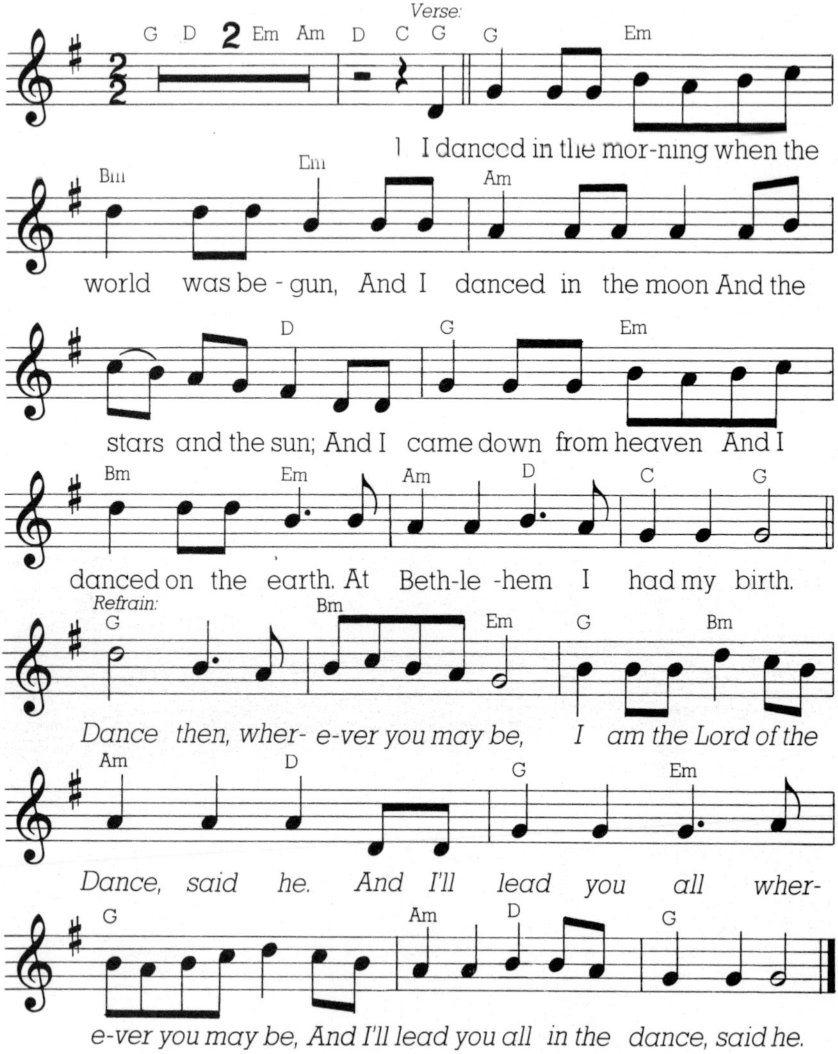
Verse:
G D 2 Em Am D C G
G Em
1 I danced in the mor-ning when the
Bm Em Am
world was be - gun, And I danced in the moon And the
D G Em
stars and the sun; And I came down from heaven And I
Bm Em Am D C G
danced on the earth. At Beth-le -hem I had my birth.
Refrain:
G Bm Em G Bm
Dance then, wher- e-ver you may be, I am the Lord of the
Am D G Em
Dance, said he. And I'll lead you all wher-
G Am D G
e-ver you may be, And I'll lead you all in the dance, said he.

2 I danced for the scribe
   And the pharisee,
   But they would not dance
   And they wouldn't follow me.
   I danced for the fishermen,
   For James and John –
   They came with me
   And the dance went on.
   Refrain:

3 I danced on the Sabbath
   And I cured the lame;
   The holy people
   Said it was a shame.
   They whipped and they stripped
   And they hung me on high,
   And they left me there
   On a Cross to die
   Refrain:

4 I danced on a Friday
   When the sky turned black;
   It's hard to dance
   With the devil on your back.
   They buried my body
   And they thought I'd gone
   But I am the dance,
   And I still go on.
   Refrain:

5 They cut me down
   And I leapt up high;
   I am the life
   That'll never, never die.
   I'll live in you
   If you'll live in me;
   I am the Lord
   Of the Dance, said he.
   Refrain:

59 Christ is the world's Light

Fred Pratt Green, b.1903

CHRISTE SANCTORUM

Melody from La Feillée *Methode du plain-chant*, 1782

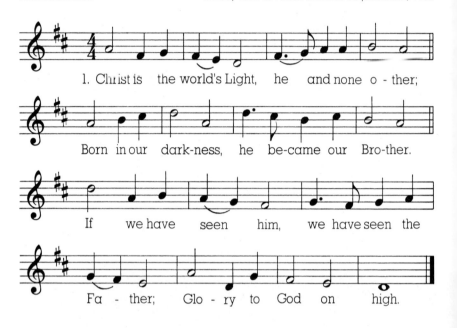

1. Christ is the world's Light, he and none o-ther;
Born in our dark-ness, he be-came our Bro-ther.
If we have seen him, we have seen the
Fa - ther; Glo - ry to God on high.

2 Christ is the world's Peace, he and none other;
No man can serve him and despise his brother.
Who else unites us, one in God the Father?
Glory to God on high.

3 Christ is the world's Life, he and none other
Sold once for silver, murdered here, our Brother –
He, who redeems us, reigns with God the Father:
Glory to God on high.

4 Give God the glory, God and none other;
Give God the glory, Spirit, Son and Father;
Give God the glory, God in man my brother;
Glory to God on high.

60 In that land which we call holy

Fred Pratt Green, b.1903
From *Psalmodia Sacra*, 1715, based on a melody by
C. F. Witt, 1660–1715

STUTTGART

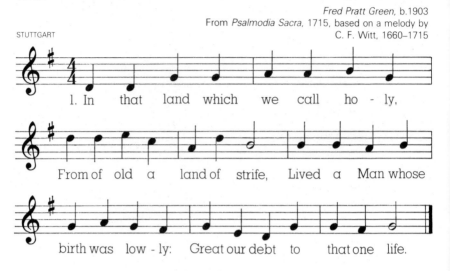

1. In that land which we call ho - ly,
From of old a land of strife, Lived a Man whose
birth was low - ly: Great our debt to that one life.

2 Where the Roman legions sweated,
In a world where might was right,
Lived a Man whose love defeated
Deadlier foes than soldiers fight.

3 Where a proud and subject nation
Learned to scorn each lesser breed,
Lived a Man whose true compassion
Knew no bounds of race or creed.

4 Where men studied to be righteous,
Strict to keep each trivial ban,
Lived a Man who came to teach us
Love of God is love of man.

5 Where God's People long expected
God would reign, or God had lied,
Lived a Man they all rejected,
Lived the God they crucified.

6 This our faith: he lives for ever!
Love redeems, though it is slain!
This his Church's whole endeavour:
So to live that Christ may reign.

The Flowering Tree

Christ looked at the people.
He saw them assailed by fear.
He saw the locked door,
He saw the knife in hand,
He saw the buried coin,
He saw the unworn coat
consumed by moth:
He saw the stagnant water
drawn and kept in the pitcher,
the musty bread in the bin,
the defended,
the unshared,
the ungiven.
He told them then,
of love that casts out fear.
Of love that is four walls,
and a roof over the head.
Of the knife in the sheath,
the coin in the open hand,
of the coat given
warm with the giver's life.
Of the water poured in the cup,
of the table spread.
The undefended,
the shared,
the given,
the Kingdom of Heaven.

He lifted his large and
beautiful hands to bless.

Caryll Houselander

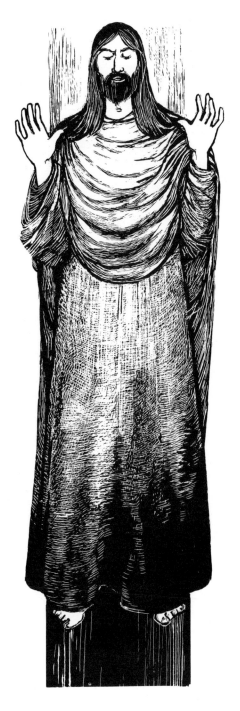

61 Jesus, good above all other

P. Dearmer, 1867–1936
Melody from a 14th-century German MS.,
QUEM PASTORES LAUDAVERE   arranged and harmonized by Ralph Vaughan Williams 1872–1958

1. Je - sus, good   a - bove all   o - ther,   Gen - tle

child of gen - tle mo-ther,   In   a   sta-ble born our

bro-ther,   Give   us   grace to   per - se - vere.

2 Jesus, cradled in a manger,
   For us facing every danger,
   Living as a homeless stranger,
       Make we thee our King most dear.

3 Jesus, for thy people dying,
   Risen Master, death defying,
   Lord in heaven, thy grace supplying,
       Keep us to thy presence near.

4 Jesus, who our sorrows bearest,
   All our thoughts and hopes thou sharest,
   Thou to man the truth declarest;
       Help us all thy truth to hear.

5 Lord, in all our doings guide us;
   Pride and hate shall ne'er divide us;
   We'll go on with thee beside us,
       And with joy we'll persevere!

You are not only good, but the cause of goodness in others.
Socrates to Protagoras in Plato's Protagoras

Matthew speaks of Jesus

I understood,
when he began to teach,
why first,
he had given light to blind eyes,
and to deaf ears,
the music of water and wind,
and to hands and feet that were numb,
the touch of delicate grass, and the sun,
and speech to the dumb . . .

Caryll Houselander

Prayer

Jesus, friend of the friendless,
Helper of the poor,
Healer of the sick,
Whose life was spent in doing good,
Let me follow in thy footsteps.
Make me loving in all my words,
And in all my deeds;
Make me strong to do right,
Gentle with the weak,
And kind to all who are in sorrow:
That I may be like thee,
My dear Lord and Master. *Amen*

'Lord, when did we do these things for you?' 'Truly, I say unto you, as
you did it to one of the humblest of these my brothers here, you did it
for me.'

Matthew 25 v 40

62 Lord of all hopefulness

Jan Struther, 1901–53
Traditional Irish, arranged by
Martin Shaw, 1875–1958

SLANE

1. Lord of all hope-ful-ness, Lord of all joy,
Whose trust, e-ver child-like, no cares could de-stroy,
Be there at our wa-king, and give us, we pray, Your
bliss in our hearts, Lord, at the break of the day.

2 Lord of all eagerness, Lord of all faith,
Whose strong hands were skilled at the plane and the lathe,
Be there at our labours, and give us, we pray,
Your strength in our hearts, Lord, at the noon of the day.

3 Lord of all kindliness, Lord of all grace,
Your hands swift to welcome, your arms to embrace,
Be there at our homing, and give us, we pray,
Your love in our hearts, Lord, at the eve of the day.

4 Lord of all gentleness, Lord of all calm,
Whose voice is contentment, whose presence is balm,
Be there at our sleeping, and give us, we pray,
Your peace in our hearts, Lord, at the end of the day.

Prayer

O Lord, bless our school:
that, working together
and playing together,
we may learn to serve you
and to serve one another:
for Jesus' sake. *Amen*

From the classroom window

Sometimes, when heads are deep in books,
And nothing stirs,
The sunlight touches that far hill,
And its three dark firs;
Then on those trees I fix my eyes –
And teacher hers.

Together awhile we contemplate
The air-blue sky
And those dark tree-tops; till, with a tiny
Start and sigh,
She turns again to the printed page –
And so do I.

But our two thoughts have met out there
Where no school is –
Where, among call of birds and faint
Shimmer of bees,
They rise in sunlight, resinous, warm –
Those dark fir-trees!

John Walsh

Prayer

Jesus, when you were on earth you had friends who were especially
close to you. You knew what it was like to enjoy their company; you
also knew what it was like when they deserted you. Please keep my
friends in your care. Help me to be a good friend. *Amen*

63 When from the sky

QUORN

Alan T. Dale, 1902–79
John Whitworth, b.1921

2 When all around us the glory of autumn
   colours the gardens, the fields and the hills,
   we sing of the wonder, unspeakable wonder,
   of God who with joy both begins and fulfils.

3 When in the coldness and deadness of winter
   storms from the east with their bluster begin,
   we sing of that morning, mysterious morning,
   when Jesus was born in the barn of an inn.

116

4 When in the gladness and greenness of springtime
    winter is over in life and in light,
    we sing of that Easter, miraculous Easter,
    that shattered the darkness and dread of the night.

64 Blest are the pure in heart

J. Keble, 1792–1866
W. H. Havergal, 1793–1870
(founded on a melody by J. B. Konig)

FRANCONIA

1. Blest are the pure in heart, For they shall see our God, The se-cret of the Lord is theirs, Their soul is Christ's a - bode.

2 The Lord, who left the heavens
Our life and peace to bring,
To dwell in lowliness with men,
Their Pattern and their King;

3 Still to the lowly soul
He doth himself impart,
And for his dwelling and his throne
Chooseth the pure in heart.

4 Lord, we thy presence seek;
May ours this blessing be;
Give us a pure and lowly heart
A temple fit for thee.

Prayers

Lord, on the way to goodness, when we stumble, hold us, when we fall, lift us up, when we are hard pressed by evil, deliver us, when we turn from what is good, turn us back, and bring us at last to your glory. *Amen*

O God, make us children of quietness and heirs of peace. *Amen*

1st-century prayer by St Clement

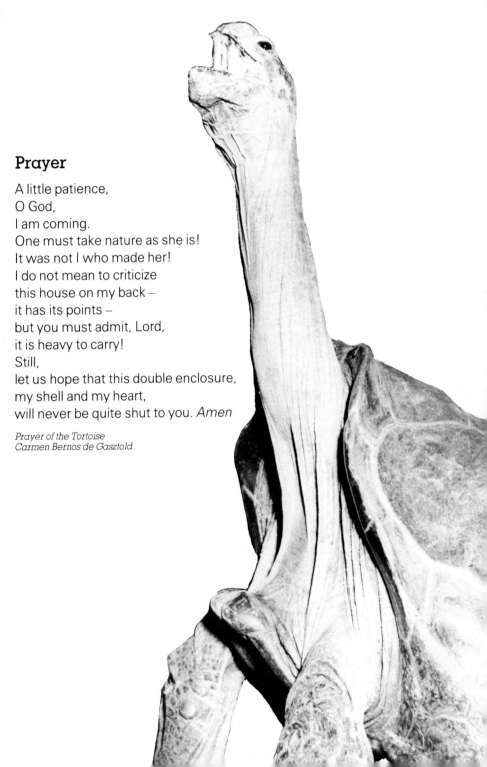

Prayer

A little patience,
O God,
I am coming.
One must take nature as she is!
It was not I who made her!
I do not mean to criticize
this house on my back –
it has its points –
but you must admit, Lord,
it is heavy to carry!
Still,
let us hope that this double enclosure,
my shell and my heart,
will never be quite shut to you. *Amen*

Prayer of the Tortoise
Carmen Bernos de Gasztold

65 Bread is the laughter

REBEKAH

Marian Collihole, b.1933
Peter D. Smith, b.1938

1. Bread is the laugh-ter of the man in the field,

Books are the laugh-ter of the wise, *Love is the laughter of the*

man who walks with God, Love is the laugh-ter of the Lord.

2 Fish are the laughter of the man in his boat,
Lambs are his laughter in the hills,
Refrain:

3 Birds are the laughter of the man in the wood,
Songs are the laughter of his soul.
Refrain:

4 A child is the laughter of the mother in her home,
Friends are the laughter of her child,
Refrain:

Prayer

O God,
Look on us
And be always with us
That we may live happily. *Amen*

A prayer of the Amazulu people

The secret of life is not to do what one likes, but to try to like what one has to do.

King George V, 1865–1936

66 Give to us eyes

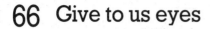

Peggy Blakeley
John Whitworth, b. 1921

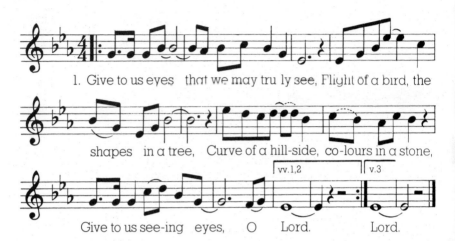

1. Give to us eyes that we may tru ly see, Flight of a bird, the shapes in a tree, Curve of a hill-side, co-lours in a stone, Give to us see-ing eyes, O Lord. Lord.

2 Give to us ears that we may truly hear,
Music in bird-song, rippling water clear,
Whine of the winter wind, laughter of a friend,
Give to us hearing ears, O Lord.

3 Give to us hands that we may truly know,
Patterns in tree bark, crispness of the snow,
Smooth feel of velvet, shapes in a shell,
Give to us knowing hands, O Lord.

I listen with reverence to the birdsong cascading
At dawn from the oasis, for it seems to me
There is no better evidence for the existence of God
Than in the bird that sings, though it knows not why,
From a spring of untrammelled joy that wells up in its heart.
Therefore I pray that no sky-hurled hawk may come
Plummeting down,
To silence the singer, and disrupt the song.

An Arab chieftain

Look at the birds flying around; they do not plant seeds, gather a harvest and put it into barns; your Father in heaven takes care of them. Aren't you worth much more than the birds?

Matthew 6 v 26

67 Father, hear the prayer we offer

Mrs L. M. Willis, 1824–1908
From a Traditional English Melody, collected, adapted, and harmonized
by Ralph Vaughan Williams, 1872–1958

SUSSEX

1. Fa - ther, hear the prayer we of - fer:
Not for ease that prayer shall be, But for strength that
we may e - ver Live our lives cou - ra - geous-ly.

2 Not for ever in green pastures
   Do we ask our way to be;
   But the steep and rugged pathway
   May we tread rejoicingly.

3 Not for ever by still waters
   Would we idly rest and stay;
   But would smite the living fountains
   From the rocks along our way.

4 Be our strength in hours of weakness;
   In our wanderings be our guide;
   Through endeavour, failure, danger,
   Father be thou at our side.

Prayer

May we show patience in everyday disappointments,
courage in everyday tasks and loyalty and understanding
in everyday friendships. *Amen*

Prayers

Lord, be thou within us, to strengthen us;
Without us, to keep us;
Above us, to protect us;
Beneath us, to uphold us;
Before us, to direct us;
Behind us, to keep us from straying;
Round about us, to defend us. *Amen*

Blessed be thou, O Lord our Father, for ever and ever.

Bishop Launcelot Andrewes, 1555–1626

Help us, O God, not to let little things get us down. Help us to sort out in our own minds what really matters and what does not, and, having done this, to act with courage, dignity and patience. *Amen*

68 Give me joy in my heart

Traditional, arranged by John Whitworth, b.1921

1. Give me joy in my heart, keep me prai-sing, Give me joy in my heart, I pray; Give me joy in my heart, keep me praising, Keep me praising till the break of day: *Sing Ho-san-na! Sing Ho-san-na! Sing Ho-san-na to the King of Kings!* King.

2 Give me peace in my heart, keep me loving,
   Give me peace in my heart, I pray,
   Give me peace in my heart, keep me loving,
   Keep me loving till the break of day:
   Refrain:

3 Give me love in my heart, keep me serving,
   Give me love in my heart, I pray,
   Give me love in my heart, keep me serving,
   Keep me serving till the break of day:
   Refrain:

Prayers

Take from us, O God,
All pride and vanity,
All boasting and forwardness,
And give us the true courage that shows itself by
    gentleness;
The true wisdom that shows itself by simplicity;
And the true power that shows itself by modesty;
Through Jesus Christ our Lord. *Amen*

Charles Kingsley, 1819–75

Teach us, good Lord, to serve thee as thou deservest;
to give and not count the cost, to fight and not heed the
wounds, to toil and not seek for rest, to labour and not
ask for any reward, save that of knowing that we do thy
will. *Amen*

St Ignatius Loyola, 1491–1556

69 Give me peace, O Lord, I pray

Estelle White
Estelle White, b.1916

1. Give me peace, O Lord, I pray, In my work and in my play, And in-side my heart and mind, Lord, give me peace.

2 Give peace to the world, I pray,
let all quarrels cease today.
May we spread our light and love.
Lord, give us peace

Let there be peace on earth, and let it begin with me.

Blessed are the peacemakers, for they shall be called the Sons of God.

Matthew 5 v 9

The wolf shall dwell with the lamb,
and the leopard shall lie down with the kid;
and the calf and the young lion
and the fatling together;
and a little child shall lead them.

Isaiah 11 v 6

70 Father, we thank thee for the night

Rebecca J. Weston, 19th century
Later form of melody by Richard Wainwright, 1758–1825,
arranged by Martin Shaw, 1875–1958

WAINWRIGHT

1. Fa - ther, we thank thee for the night,
And for the plea-sant mor-ning light; For rest and food and
lo-ving care, And all that makes the day so fair.

2 Help us to do the things we should,
   To be to others kind and good;
   In all we do at work or play
   To grow more loving every day.

Prayer

O Father God, help us to do our best at work and play.
Make us adventurous in tackling the tasks we find
difficult. Teach us to be eager to help at home and at
school, and show us how to share our joy and gladness
with other people. *Amen*

The quarrel

I quarrelled with my brother
I don't know what about,
One thing led to another
And somehow we fell out.
The start of it was slight,
The end of it was strong.
He said he was right,
I knew he was wrong!

We hated one another,
The afternoon turned black.
Then suddenly my brother
Thumped me on the back,
And said, 'Oh come along!
We can't go on all night –
I was in the wrong.'
So he was in the right.

Eleanor Farjeon

He who forgives, ends the quarrel.

African proverb

71 God, whose farm is all creation

John Arlott, b.1914
Traditional English, arranged and harmonized
by Ralph Vaughan Williams, 1872–1958

SHIPSTON

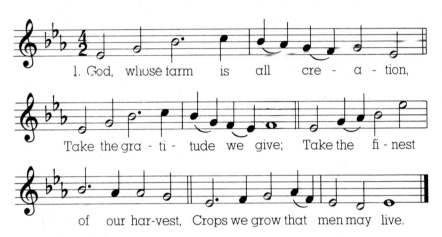

1. God, whose farm is all cre - a - tion,
Take the gra - ti - tude we give; Take the fi - nest
of our har-vest, Crops we grow that men may live.

2 Take our ploughing, seeding, reaping,
   Hopes and fears of sun and rain,
   All our thinking, planning, waiting,
   Ripened in this fruit and grain.

3 All our labour, all our watching,
   All our calendar of care,
   In these crops of your creation,
   Take, O God: they are our prayer.

Prayer

Dear God, there are times when we try very hard at a task and get
nowhere. There are times when other people understand and learn
easily and we fail. There are times when others race on and we are left
behind. Give us patience then, O God, to persevere and try even
harder, rather than give up. *Amen*

Bread

Be gentle when you touch Bread.
Let it not lie
Uncared for,
Unwanted.
So often Bread
Is taken for granted.

There is such beauty in Bread.
Beauty of sun and soil,
Beauty of patient toil.
Wind and rain
Have caressed it.
Christ often blessed it.
Be gentle when you touch Bread.

Freda Elton Young

72 God be in my head

DAVID

Sarum Primer, 1558
G. W. Briggs, 1875–1959

God be in my head, And in my un-der - stan-ding;

God be in mine eyes, And in my loo - king;

God be in my mouth and in my spea - king;

God be in my heart, And in my thin - king.

God be at mine end. And at my de - par - ting.

Prayers

Dear Lord Jesus,
we shall have this day only once;
before it is gone,
help us to do all the good we can,
so that today is not a wasted day.
For thy name's sake. *Amen*

*Freely adapted from a prayer attributed to
Stephen Grellet, 1773–1855*

O God help us not to despise or oppose what we do not understand. *Ame*

William Penn, Quaker, 1644–1718

134

All but blind

All but blind
  In his chambered hole
Gropes for worms
  The four-clawed Mole.

All but blind
  In the evening sky
The hooded Bat
  Twirls softly by.

All but blind
  In the burning day
The Barn-Owl blunders
  On her way.

And blind as are
  These three to me,
So, blind to Some-One
  I must be.

Walter de la Mare

73 Grant us your peace, Lord

Francesca Leftley
Israeli Folk Song, arranged by Francesca Leftley, b.1955

1. Grant us your peace, Lord, shelter us from harm, Lord, grant us your peace, Lord, shield us with your love. Just as a father cares for his children, grant us your peace, Lord, shield us with your love.

2 Grant us your strength, Lord,
shelter us from harm, Lord,
grant us your strength, Lord,
shield us with your love.
From dusk till daybreak, each hour of each day,
grant us your strength, Lord,
shield us with your love.

Prayer

O God, you have made me, and you will keep me. I am never alone:
for you are always by my side, and I am safe with you. Help me to fear
nothing: but always to trust you and to be brave. *Amen*

God shall judge between nations;
they shall beat their swords into ploughshares,
and their spears into pruning hooks;
nation shall not lift up sword against nation;
neither shall they learn war any more.

Isaiah 2 v 4

Jesus said, 'Peace I leave with you, my peace I give to you.'

John 14 v 27

74 In Christ there is no East or West

John Oxenham, 1852–1941
Scottish Psalter, 1615, as given in
Ravencroft's Psalter, 1621

DUNDEE

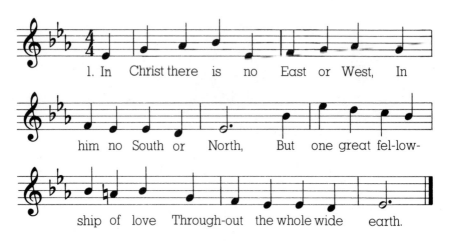

1. In Christ there is no East or West, In him no South or North, But one great fel-low-ship of love Through-out the whole wide earth.

2 In him shall true hearts everywhere
   Their high communion find,
   His service is the golden cord
   Close-binding all mankind.

3 Join hands, then, brothers of the Faith,
   What'er your race may be;
   Who serves my Father as a son
   Is surely kin to me.

4 In Christ now meet both East and West,
   In him meet South and North,
   All Christlike souls are one in him,
   Throughout the whole wide earth.

Prayer

Help us, O God, to appreciate those who are different from ourselves. Give us the chance to learn from those who choose different games, books and pastimes, and from those whose beliefs and ways of life are unusual to us. Help us to see the best in all people and to look on these differences as an enrichment of our own lives. *Amen*

Interesting people are those who take an interest in all manner of things.

There are no rich, no poor, no black, no white before God. It is your actions that make you good or bad.

Guru Nanak

No one is born prejudiced against others, but everyone is born prejudiced in favour of himself.

Dr David Stafford Clark

75 Heavenly Father

W. C. Piggott, 1872–1943
R. H. Prichard, 1811–87

HYFRYDOL

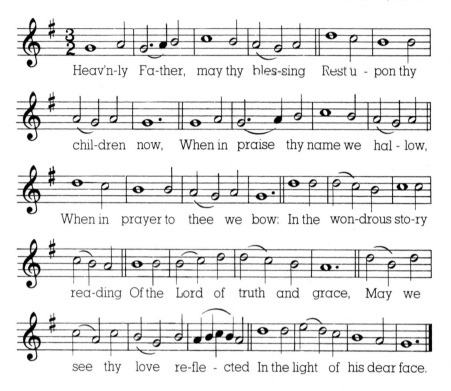

Heav'n-ly Fa-ther, may thy bles-sing Rest u - pon thy chil-dren now, When in praise thy name we hal - low, When in prayer to thee we bow: In the won-drous sto-ry rea-ding Of the Lord of truth and grace, May we see thy love re-fle - cted In the light of his dear face.

2 May we learn from his great story
    All the arts of friendliness;
Truthful speech and honest action,
    Courage, patience, steadfastness;
How to master self and temper,
    How to make our conduct fair;
When to speak and when be silent,
    When to do and when forbear.

3 May his spirit wise and holy
    With his gifts their spirits bless,
Make us loving, joyous, peaceful,
    Rich in goodness, gentleness,
Strong in self-control, and faithful,
    Kind in thought and deed; for he
Sayeth, 'What ye do for others
    Ye are doing unto me.'

Jesus said, 'I was hungry and you gave me food, I was thirsty and you gave me drink, I was a stranger and you welcomed me, I was naked and you clothed me, I was sick and you visited me, I was in prison and you came to me.'

Matthew 25 vv 35–37

If you want a friend, be a friend.

76 Kum ba yah

Arranged by John Whitworth, b.1921

1. Kum ba yah, my Lord, kum ba yah, Kum ba
yah, my Lord, kum ba yah, Kum ba yah, my Lord, kum ba
yah, O Lord, kum ba yah. Kum ba yah.

2 Someone's singing, Lord, kum ba yah.

3 Someone's praying, Lord, kum ba yah.

4 Someone's hungry, Lord, kum ba yah.

5 Someone's suffering, Lord, kum ba yah.

6 Someone's lonely, Lord, kum ba yah.

Prayer

O God, help us to remember those who find everyday tasks difficult.
Help us to remember the blind and the deaf and those whose limbs
are awkward or useless. May we be considerate and caring at all times
and may we take example from the courage they display in performing
day to day tasks. *Amen*

Prayer

Dear God, we shall not always be healthy and happy. Help us at times of illness to be brave and co-operative with those who try to help us. And at times of disappointment with others, give us patience and understanding and a willingness to forgive. *Amen*

Ask, and it shall be given you; seek, and ye shall find; knock, and it shall be opened to you.

Matthew 7 v 7

77 Life is great

LITHEROP

Brian Wren, b.1936
P. Cutts, b.1937

1. Life is great! So sing a-bout it, as we can and as we should shops and bu -ses, towns and peo-ple, vil-lage, farm-lands, field and wood. Life is great and life is gi-ven.

vv.1,2,3 / v.4

Life is love-ly, free and good. li - ving Man.

2 Life is great! – whatever happens,
snow or sunshine, joy or pain,
hardship, grief, or disillusion,
suffering that I can't explain –
Life is great if someone loves me,
holds my hand and calls my name.

3 Love is great! – the love of lovers,
whispered words and longing eyes;
love that gazes at the cradle
where a child of loving lies;
love that lasts when youth has faded,
bonds with age, but never dies.

4 God is great! In Christ he loved us,
as we should but never can –
love that suffered, hoped and trusted,
when disciples turned and ran,
love that broke through death for ever –
Praise that loving, living Man.

Danny Murphy

He was as old as old could be,
His little eye could scarcely see,
His mouth was sunken in between
His nose and chin, and he was lean
And twisted up and withered quite,
So he could hardly walk aright.

His pipe was always going out,
And then he'd have to search about
In all his pockets, and he'd mow
– Oh, deary me! and, musha now! –
And then he'd light his pipe, and then
He'd let it go clean out again.

He couldn't dance or jump or run,
Or ever have a bit of fun
Like me and Susan, when we shout
And jump and throw ourselves about:
– But when he laughed, then you could see
He was as young as young could be!

James Stephens

78 Shalom chaverim

Michael Lehr
Traditional Hebrew tune,
arranged by Michael Metcalfe, b. 1937

SHALOM

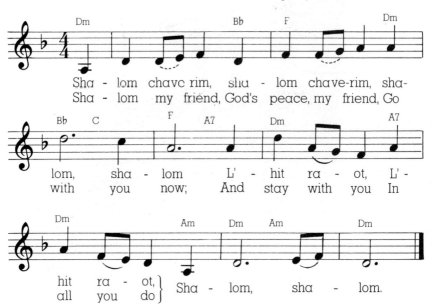

Sha - lom chave rim, sha - lom chave-rim, sha-
Sha - lom my friend, God's peace, my friend, Go

lom, sha - lom L' - hit ra - ot, L' -
with you now; And stay with you In

hit ra - ot, } Sha - lom, sha - lom.
all you do }

[Note: When singing this hymn as a round, singers may enter after every bar,
making a round in 8 parts, or after every 2 bars (4 parts) or after
4 bars (2 parts). Only the chord of Dm should be used.]

Prayers

Lord, when we are wrong, make us willing to change, and when we
are right, make us easy to live with. *Amen*

A friend is, one might say, a second self.

Cicero

Heavenly Father, be with us during this day and give us the spirit of
truth and courage and kindness. *Amen*

Friends

I fear it's very wrong of me
And yet I must admit,
When someone offers friendship
I want the *whole* of it.
I don't want everybody else
To share my friends with me.
At least, I want *one* special one,
Who, indisputably,

Likes me much more than all the rest,
Who's always on my side,
Who never cares what others say,
Who lets me come and hide
Within his shadow, in his house –
It doesn't matter where –
Who lets me simply be myself,
Who's always, *always* there.

Elizabeth Jennings

79 One more step

SOUTHCOTT

Sydney Carter
Sydney Carter, b.1915

1. One more step a-long the world I go, One more step a-long the

world I go. From the old things to the new

Keep me tra-vel-ling a-long with you. *And it's from the old I*

tra-vel to the new. Keep me tra-vel-ling a-long with you.

2 Round the corners of the world I turn,
More and more about the world I learn.
All the new things that I see
You'll be looking at along with me.
Refrain:

3 As I travel through the bad and good
Keep me travelling the way I should.
Where I see no way to go
You'll be telling me the way, I know.
Refrain:

4 Give me courage when the world is rough,
Keep me loving though the world is tough.
Leap and sing in all I do,
Keep me travelling along with you.
Refrain:

Prayer

When, as a child, I laughed and wept, time crept.
When, as a youth, I dreamed and talked, time walked.
When I became a full-grown man, time ran.
And later, as I older grew, time flew.
Soon I shall find, while travelling on, time gone.
Will Christ have saved my soul by then? *Amen*

A grandfather clock, Chester Cathedral

If you stand thinking about every step you take, you'll spend your life
standing on one foot.

Chinese proverb

80 The Lord's my shepherd

Scottish Psalter, 1650, based on Psalm 23
Melody by J. S. Irvine, 1836–87,
arranged by T. C. L. Pritchard

CRIMOND

1. The Lord's my Shep - herd, I'll not want;
He makes me down to lie
In pa - stures green; He lea - deth me
The qui - et wa - ters by.

[Note: The guitar chords should not be used with the keyboard accompaniment.]

2 My soul he doth restore again;
And me to walk doth make
Within the paths of righteousness
E'en for his own name's sake.

3 Yea, though I walk in death's dark vale,
Yet will I fear none ill:
For thou art with me; and thy rod
And staff me comfort still.

4 My table thou hast furnished
In presence of my foes;
My head thou dost with oil anoin
And my cup overflows.

5 Goodness and mercy all my life
Shall surely follow me;
And in God's house for evermore
My dwelling place shall be.

Prayer

Jesus, you understand my fears better than I. Let me trust in you and
let me help others to face their fears as you help me face mine. *Amen*

God is our refuge and strength, a very present help in trouble.

Psalm 46 v 1

The Lord is my shepherd, I shall not want;
He makes me to lie down in green pastures.
He leads me beside still waters;
He restores my soul.
He leads me in the paths of righteousness, for his name's sake.
Even though I walk in the valley of the shadow of death
I shall feel no evil;
For thou art with me; thy rod and thy staff
They comfort me.
Thou preparest a table before me in the presence of mine enemies;
Thou anointest my head with oil, my cup overflows.
Surely goodness and mercy shall follow me all the days of my life;
and I shall dwell in the house of the Lord for ever.

Psalm 23

81 When a knight won his spurs

Jan Struther, 1901–53
Adapted from an English Traditional Melody, arranged and harmonized
by Ralph Vaughan Williams, 1872–1958

STOWEY

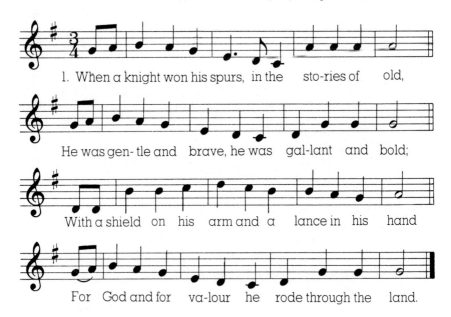

1. When a knight won his spurs, in the sto-ries of old,

He was gen-tle and brave, he was gal-lant and bold;

With a shield on his arm and a lance in his hand

For God and for va-lour he rode through the land.

2 No charger have I, and no sword by my side,
Yet still to adventure and battle I ride,
Though back into storyland giants have fled,
And the knights are no more and the dragons are dead.

3 Let faith be my shield and let joy be my steed
'Gainst the dragons of anger, the ogres of greed;
And let me set free, with the sword of my youth,
From the castle of darkness the power of the truth.

Prayer

Into thy hands, O God, we commend ourselves this
day; let thy presence be with us to its close. Enable us
to feel that in doing our work we are doing thy will, and
that in serving others we are serving thee; through
Jesus Christ our Lord. *Amen*

Prayers

Let this day, O Lord, add some knowledge or good deed to yesterday.
Amen

Bishop Launcelot Andrewes, 1555–1626

God grant us serenity to accept the things we cannot change, courage
to change the things we can, and wisdom to know the difference.

Victor Gollancz

82 When I needed a neighbour

Sydney Carter
Sydney Carter

1. When I nee-ded a neighbour, were you

there, were you there? When I nee-ded a neigh-bour, were you

there? *And the creed and the co-lour and the*

name won't mat-ter, Were you there?

2 I was hungry and thirsty, were you there, were you there?
I was hungry and thirsty, were you there?
Refrain:

3 When I needed a shelter, were you there, were you there?
When I needed a shelter, were you there?
Refrain:

4 Wherever you travel, I'll be there, I'll be there,
Wherever you travel, I'll be there.
*And the creed and the colour and the name won't matter.
I'll be there.*

Be at peace with one another.

Mark 9 v 50

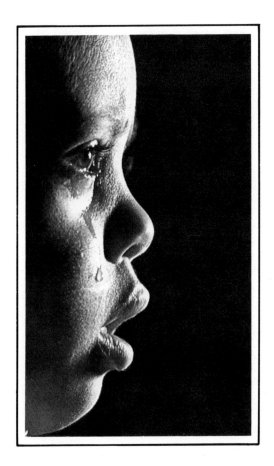

The Prophet said, 'He is not a believer who eats his fill when his neighbour beside him is hungry.'

Always treat others as you would like them to treat you.

Matthew 7 v 12

Prayer

Hear our prayer, O God, for the hungry and starving people of the world; for the homeless and the refugee; for the sick and diseased and for all who are in distress. Bless all who make themselves neighbours to such as these – the relief organizations and all doctors, nurses, teachers, scientists and others who have readily offered their services. We ask it in the name of Jesus Christ our Lord. *Amen*

83 God give me ears

Translated by Jacqueline Froom
Old Icelandic Melody, arranged by Kenneth Pont

ICELAND

1. God, give me ears that will al - ways hear thee; God, give me eyes that will see far and wide. God, give me hands that will bring beau-ty near me; God, give me feet that will walk by thy side. God, give me ears that will always hear thee; God, give me eyes that will see far and wide.

2 God, fill my thoughts with the spirit of seeking,
  Visions of wonder let me ever see.
  Give to my tongue only truth when I'm speaking,
  Fill my heart with the knowledge of thee.
  God, fill my thoughts with the spirit of seeking,
  Visions of wonder let me ever see.

156

Have you not heard?

Have you not heard his silent steps?
He comes, comes, ever comes,
Every moment, and every age,
Every night and every day,
He comes, comes, ever comes.

Rabindranath Tagore

The world around us

84 All things bright and beautiful

Mrs C. F. Alexander, 1818–95
Traditional English, arranged by
Martin Shaw, 1875–1958

ROYAL OAK

Refrain:

1. All things bright and beau-ti-ful, All crea tures great and small, All things wise and won - der - ful, The Lord God made them all.

2. Each lit - tle flower that o-pens, Each lit - tle bird that sings, He made their glowing co - lours, He made their ti - ny wings:

3 The purple-headed mountain,
    The river running by,
The sunset and the morning,
    That brightens up the sky: *Refrain:*

4 The cold wind in the winter,
    The pleasant summer sun,
The ripe fruits in the garden,
    He made them every one: *Refrain:*

5 He gave us eyes to see them,
    And lips that we might tell
How great is God Almighty,
    Who has made all things well: *Refrain:*

Tall nettles

Tall nettles cover up, as they have done
These many springs, the rusty harrow, the plough
Long worn out, and the roller made of stone:
Only the elm butt tops the nettles now.

This corner of the farmyard I like most:
As well as any bloom upon a flower
I like the dust on the nettles, never lost
Except to prove the sweetness of a shower.

Edward Thomas

Only those who love God's creation will find him.

Guru Govind Singh

Prayer

O heavenly Father, protect and bless all things that have breath: guard
them from all evil and let them sleep in peace. *Amen*

Albert Schweitzer

85 Come, ye thankful people, come

ST GEORGE

H. *Alford*, 1810–71
G. J. Elvey, 1816–93

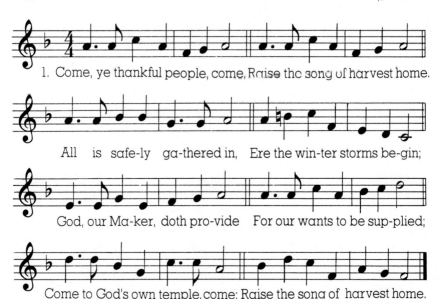

1. Come, ye thankful people, come, Raise the song of harvest home.

All is safe-ly ga-thered in, Ere the win-ter storms be-gin;

God, our Ma-ker, doth pro-vide For our wants to be sup-plied;

Come to God's own temple, come; Raise the song of harvest home.

2 All this world is God's own field,
Fruit unto his praise to yield;
Wheat and tares together sown,
Unto joy or sorrow grown;
First the blade and then the ear,
Then the full corn shall appear;
Lord of harvest, grant that we
Wholesome grain and pure may be.

The desert shall rejoice and blossom as a rose.

Isaiah 35 v 1

The fruit of the spirit of God is love, joy, peace, patience, kindness,
goodness, faithfulness, gentleness, self-control.

Galatians 5 vv 22–23

Prayer

O God, giver of the seed, the rain, the sun and the
harvest, we thank you for all who have learned to
obtain better harvests from the land, in the orchard, for
the desert, or from the sea. Bless all who work for
better harvests where modern methods are unknown
and people starve. May those who have money and
knowledge help those who have not; in the name of
Jesus Christ. *Amen*

While the earth remains, seedtime and harvest, cold and heat,
summer and winter, day and night shall not cease.

I set my bow in the clouds, and it shall be the token of a covenant
which I have established between me and all the people on the earth.

Genesis 8 v 22, 9 v 13

86 Glad that I live am I

L. W. Reese, 1856–1935
Geoffrey Shaw, 1879–1943

WATER-END

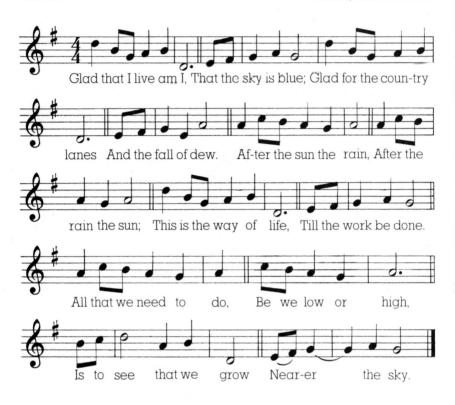

Glad that I live am I, That the sky is blue; Glad for the coun-try

lanes And the fall of dew.   Af-ter the sun the rain, After the

rain the sun; This is the way of life, Till the work be done.

All that we need to   do,   Be we low or   high,

Is to see   that we   grow Near-er   the sky.

maggie and milly and molly and may

maggie and milly and molly and may
went down to the beach (to play one day)

and maggie discovered a shell that sang
so sweetly she couldn't remember her troubles, and

milly befriended a stranded star
whose rays five languid fingers were;

and molly was chased by a horrible thing
which raced sideways while blowing bubbles; and

may came home with a smooth round stone
as small as a world and as large as alone.

for whatever we lose (like a you or a me)
it's always ourselves we find in the sea

e. e. cummings

163

87 Far round the world

WOODLANDS

Basil Mathews, 1879–1951
W. Greatorex, 1877–1949

1. Far round the world thy chil-dren sing their song;
From East and West their voi-ces sweet-ly blend,
Pra-sing the Lord in whom young lives are strong,
Je - sus our guide, our he-ro and our friend.

2 Guide of the pilgrim clambering to the height,
  Hero on whom our fearful hearts depend,
  Friend of the wanderer yearning for the light,
  Jesus our guide, our hero, and our friend.

3 Where thy wide ocean, wave on rolling wave,
  Beats through the ages on each island shore,
  They praise their Lord, whose hand alone can save,
  Whose sea of love surrounds them evermore.

4 All round the world let children sing thy song,
  From East and West their voices sweetly blend;
  Praising the Lord in whom young lives are strong,
  Jesus our guide, our hero, and our friend.

All men are the same, although they appear different; The light and the dark, the ugly and the beautiful. All human beings are reflections of one and the same Lord. Recognize the whole human race as one.

Guru Govind Singh

88 For the fruits of his creation

STOKEWOOD

Fred Pratt Green, b.1903
T. Brian Coleman, b.1920

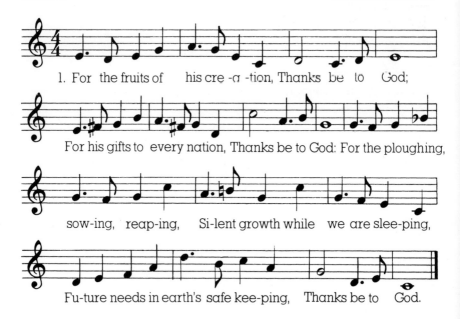

1. For the fruits of his cre-a-tion, Thanks be to God;
For his gifts to every nation, Thanks be to God: For the ploughing,
sow-ing, reap-ing, Si-lent growth while we are slee-ping,
Fu-ture needs in earth's safe kee-ping, Thanks be to God.

2 In the just reward of labour,
   God's will is done;
   In the help we give our neighbour,
   God's will is done:
   In our world-wide task of caring
   For the hungry and despairing,
   In the harvests we are sharing,
   God's will is done.

3 For the harvests of his Spirit,
   Thanks be to God;
   For the good we all inherit,
   Thanks be to God:
   For the wonders that astound us,
   For all truths that still confound us,
   Most of all that love has found us.
   Thanks be to God.

166

Prayers

Heavenly Father, bless us,
And keep us all alive;
There's ten of us to dinner
And not enough for five. *Amen*

O God, we ask that as we thank you for our daily bread
you will help us to supply the needs of those who are
hungry. *Amen*

John S. Williams

There is a sufficiency in the world for man's need, but not
for man's greed.

Mahatma Gandhi

89 For the beauty of the earth

F. S. Pierpoint, 1835–1917
adapted and harmonized by Geoffrey Shaw, 1879–1943

ENGLAND'S LANE

1. For the beau-ty of the earth, For the beau ty of the skies, For the love which from our birth O-ver and a- round us lies: *Fa- ther, un - to thee we raise This our joy - ful hymn of praise.*

2 For the beauty of each hour
   Of the day and of the night,
Hill and vale, and tree and flower,
   Sun and moon and stars of light:
Refrain:

3 For the joy of ear and eye,
   For the heart and brain's delight,
For the mystic harmony
   Linking sense to sound and sight:
Refrain:

4 For the joy of human love,
   Brother, sister, parent, child,
Friends on earth, and friends above
   For all gentle thoughts and mild:
Refrain:

Prayer

Most wonderful and loving God
we praise you
we worship you
we thank you
for everything you have given us.

We thank you for life itself
and for everything that makes life worth living:
for the things we like to eat,
for the things we like to do,
for the people who love us and care for us.

But as we say thank-you now
we know that we do not always say thank-you
in the way we live our lives.

We are sorry as we remember
the wrong things we have done,
the things we should not have said,
and the ways we have hurt other people. *Amen*

169

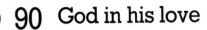

90 God in his love

STEWARDSHIP

Fred Pratt Green, b.1903
Valerie Ruddle, b.1932

1. God in his love for us lent us this pla-net,
Gave it a pur-pose in time and in space: Small as a
spark from the fire of cre - a - tion, Cra-dle of
life and the home of our race, world with-out end!

2 Thanks be to God for its bounty and beauty,
Life that sustains us in body and mind:
Plenty for all, if we learn how to share it,
Riches undreamed of to fathom and find.

3 Long have the wars of man ruined its harvest;
Long has earth bowed to the terror of force;
Long have we wasted what others have need of,
Poisoned the fountain of life at its source.

4 Earth is the Lord's: it is ours to enjoy it,
Ours, as his stewards, to farm and defend.
From its pollution, misuse, and destruction,
Good Lord deliver us, world without end!

170

Miracles

Why, who makes much of a miracle?
As to me I know of nothing else but miracles,
Whether I walk in the streets of Manhattan,
Or dart my sight over the roofs of houses towards the sky,
Or wade with naked feet along the beach just in the edge of the water,
Or stand under trees in the woods,
Or talk by day with anyone I love,
Or sit at table at dinner with the rest,
Or look at strangers riding opposite me in the car,
Or watch honey-bees busy round the hive of a summer fore-noon,
Or animals feeding in the fields,
Or birds, or the wonderfulness of insects in the air,
Or the wonderfulness of the sundown,
Or of stars shining so quiet and bright,
Or the exquisite delicate thin curve of the new moon in spring:
These with the rest, one and all, are to me miracles . . .

Walt Whitman

91 Morning has broken

Eleanor Farjeon, 1881–1965
Old Gaelic Melody,
arranged by John Whitworth, b.1921

BUNESSAN

Brightly

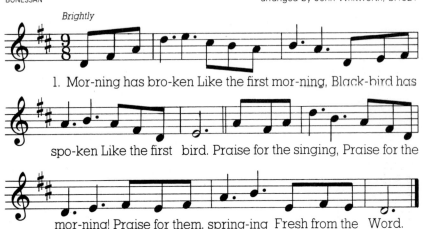

1. Mor-ning has bro-ken Like the first mor-ning, Black-bird has spo-ken Like the first bird. Praise for the singing, Praise for the mor-ning! Praise for them, spring-ing Fresh from the Word.

2 Sweet the rain's new fall
  Sunlit from heaven,
  Like the first dewfall
    On the first grass.
  Praise for the sweetness
  Of the wet garden,
  Sprung in completeness
    Where his feet pass.

3 Mine is the sunlight
  Mine is the morning
  Born of the one light
    Eden saw play.
  Praise with elation,
  Praise every morning,
  God's re-creation
    Of the new day.

Prayer

Lord of colour, we greet thee,
Lord of pattern, we rejoice in thee,
Lord of variety, we delight in thee,
Lord of life, we adore thee. *Amen*

92 Skipping down the pavement

Donald H. Hilton, b.1932
Yvonne Gooding

PATTERNS

1. Skip-ping down the pave-ment wide, count the pa-ving stones;

big or small or square or round; see their patterns on the ground.

House and factory, church and shop, bricks and stones reach high.

Bu-sy work-men made them all: see their pat-terns on the wall.

2 Look above the rooftops tall,
  far as you can see.
Clouds in daytime; stars at night;
  see their patterns in the sky.
People walking down the street,
  dressed in colours gay;
stripes and circles, frills and bows:
  see the patterns in their clothes.

3 Clouds and sunshine, night-time stars,
  clothes and curtains too;
stones and pavement, bricks and wall:
  see the patterns in them all.
Thank you, God, for sights to see
  round us every day:
still or moving; big or small;
  and the patterns in them all!

174

Prayer

For all the children who are handicapped by hard
circumstances; for those who are being brought up in
homes in which there is no beauty or joy or love; for
those who suffer through sickness, fear or neglect; we
beseech thee to hear us, O God; and grant, we pray thee,
to all who tend and teach them, patience and wisdom, for
the sake of Jesus Christ. *Amen*

93 Thank you, Lord

+ *Brian A. Wren, b.1936*
Melody by Louis Bourgeois, c.1510–c.1561

OLD 124th

1. Thank you, Lord, for wa-ter, soil and air Large gifts sup-
por-ting e-very-thing that lives. For-give our spoi-ling
and a-buse of them. Help us re-new the face of the earth,
Help us re-new the face of the earth.

2 Thank you, Lord, for minerals and ores –
   The basis of all building, wealth and speed.
   Forgive our reckless plundering and waste.
   Help us renew the face of the earth,
   Help us renew the face of the earth.

3 Thank you, Lord, for priceless energy –
   Stored in each atom, gathered from the sun.
   Forgive our greed and carelessness of power.
   Help us renew the face of the earth,
   Help us renew the face of the earth.

4 Thank you, Lord, for making planet Earth
   A home for us and ages yet unborn.
   Help us to share, consider, save and store.
   Come and renew the face of the earth,
   Come and renew the face of the earth.

Prayer

We are glad for the hardness of rock, Father,
   For its angles and edges,
   For its delicate shades of colour,
   For its grandeur in mountains,
   For its usefulness in buildings,
   For its enduring beauty in sculpture,
We are glad for the world of rock, Father.

We are glad for the fertility of soil, Father,
   For its texture and colour,
   For its fragrance after rain,
   For the life within it,
   For prairies and cornfields, meadows and gardens,
   For our own garden and window-box which we can cultivate.
We are glad for the earth, Father. *Amen*

94 Think of a world

WESTCOTT

Doreen Newport
Peter Cutts, b.1937

1. Think of a world with-out a-ny flo-wers,

Think of a world with-out a-ny trees, Think of a sky with-

out a-ny sun-shine, Think of the air with-out a-ny breeze. We

thank you, Lord, for flowers and trees and sun-shine, We

thank you, Lord, and praise your ho-ly name. name.

2 Think of the world without any animals,
   Think of a field without any herd,
   Think of a stream without any fishes,
   Think of a dawn without any bird.
   We thank you, Lord,
       for all your living creatures,
   We thank you, Lord,
       and praise your holy name.

3 Think of a world without any people,
  Think of a street with no one living there,
  Think of a town without any houses,
  No one to love and nobody to care.
  We thank you, Lord,
      for families and friendships,
  We thank you, Lord,
      and praise your holy name.

Four little foxes

Speak gently, Spring, and make no sudden sound,
For in my windy valley yesterday I found
New born foxes squirming on the ground –
Speak gently.

Walk softly, March, forbear the bitter blow.
Her feet within a trap, but blood upon the snow,
The four little foxes saw their mother go –
Walk softly.

Go lightly, Spring, oh give them no alarm;
As I covered them with boughs to shelter them from harm,
The thin blue foxes suckled at my arm –
Go lightly.

Step softly, March, with your rampant hurricane;
Nuzzling one another and whimpering with pain,
The new little foxes are shivering in the rain –
Step softly.

Lew Sarett

95 This is a lovely world

Jane Palmer
Jane Palmer, b.1952

1. This is a love - ly world. Birds in the trees a - bove Sing of a world that's made By a God of Love.

2 This is a joyful world
   Where every girl and boy
   Sings of a world that's made
   By a God of joy.

I am not interested in moderate honesty.
Who wants to draw most of his salary?
To eat an egg that is moderately fresh?
To live in a house that keeps out most of the rain?
To travel in a ship that floats most of the time?
The kind of honesty I am interested in
is absolute honesty.

Daw Nyein Tha

The thrush's nest

Within a thick and spreading hawthorn bush,
That overhung a mole-hill large and round,
I heard from morn to morn a merry thrush
Sing hymns to sunrise, and I drank the sound
With joy; and, often an intruding guest,
I watched her secret toils from day to day –
How true she warped the moss to form a nest,
And modelled it within with wood and clay;
And by and by, like heath-bells gilt with dew,
There lay her shining eggs, as bright as flowers,
Ink-spotted over shells of greeny blue;
And there I witnessed, in the sunny hours
A brood of nature's minstrels chirp and fly,
Glad as that sunshine and the laughing sky.

John Clare

96 We plough the fields

M. Claudius, 1740–1815,
translated by J. M. Campbell, 1817–78
J. A. P. Schulz, 1747–1800

WIR PFLUGEN

1. We plough the fields, and scat-ter The good seed on the land, But it is fed and wa-tered By God's al-mighty hand: He sends the snow in win-ter, The warmth to swell the grain, The breez-es and the sun-shine, And soft re-fre-shing rain. *All good gifts a-round us Are sent from heav'n a-bove; Then thank the Lord, O thank the Lord For all his love.*

2 He only is the Maker
   Of all things near and far,
He paints the wayside flower,
   He lights the evening star.
The winds and waves obey him,
   By him the birds are fed;
Much more to us, his children,
   He gives our daily bread.
Refrain:

3 We thank thee then, O Father,
   For all things bright and good;
The seed-time and the harvest,
   Our life, our health, our food.
No gifts have we to offer
   For all thy love imparts,
But that which thou desirest,
   Our humble, thankful hearts.
Refrain:

For everything there is a season,
and a time for every matter under heaven:
a time to be born, and a time to die;
a time to plant, and a time to pluck up what is planted.

Ecclesiastes 3 vv 1–2

Prayer

Blessed art thou,
O Lord our God,
king of the universe,
who bringest forth bread from the earth. *Amen*

Jewish blessing

97 When lamps are lighted

M. M. Penstone, 1859–1910
Traditional English, arranged by
Martin Shaw, 1875–1958

BUTLER

1. When lamps are ligh-ted in the town, The
boats sail out to sea; The fi-shers watch when
night comes down, They work for you and me.

2 When little children go to rest,
    Before they sleep, they pray
  That God will bless the fishermen
    And bring them back at day.

3 The boats come in at early dawn,
    When children wake in bed;
  Upon the beach the boats are drawn,
    And all the nets are spread.

4 God hath watched o'er the fishermen
    Far on the deep dark sea,
  And brought them safely home again,
    Where they are glad to be.

Until I saw the sea

Until I saw the sea
I did not know
that wind
could wrinkle water so.

I never knew
that sun
could splinter a whole sea of blue.

Nor
did I know before,
a sea breathes in and out
upon a shore.

Lilian Moore

98 National Anthem

Thesaurus Musicus, c.1743

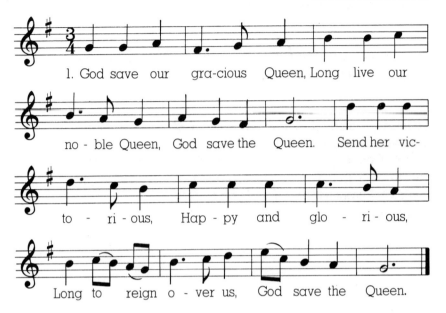

1. God save our gra-cious Queen, Long live our no-ble Queen, God save the Queen. Send her vic-to-ri-ous, Hap-py and glo-ri-ous, Long to reign o-ver us, God save the Queen.

2 Thy choicest gifts in store
   On her be pleased to pour,
   Long may she reign!
   May she defend our laws,
   And ever give us cause
   To sing with heart and voice,
   God save the Queen.

Prayers

Let us pray for our country: there is much wrong with it, but much more to be thankful for. Let us pray for those who do their best to govern our country: they often make mistakes but they are capable of doing much good. Now let us pray for ourselves: this country is our home; we have the power to spoil it and waste its resources or to help make it a better place for us all. *Amen*

Dear God, it is better to have friends than enemies. We shall learn in history that, long ago, our nation had many enemies in Europe – French, Germans, Dutch, Spanish and Italians. Now we have learnt how to make friends with our neighbours. We pray for our good friends in Europe with whom we have much to share and much to enjoy. *Amen*

Index of first lines

Hymn

A is for Advent 24
All creatures of our God and King 1
Alleluia 2
All glory, laud and honour 43
All people that on earth do dwell 3
All things bright and beautiful 84
Away in a manger, no crib for a bed 29

Blest are the pure in heart 64
Bread is the laughter of the man in the field 65

Christ is the world's Light, he and none other 59
Come to us, creative Spirit 54
Come, ye thankful people, come 85

Every star shall sing a carol 28
Everything changes 4

Far round the world thy children sing their song 87
Father, hear the prayer we offer 67
Father, we adore you 5
Father, we thank thee for the night 70
Father, we thank you 6
Fire is lighting torch and lamp at night 55
For all the love that from our earliest days 8
For the beauty of the earth 89
For the fruits of his creation 88
Forty days and forty nights 42

Give me joy in my heart, keep me praising 68
Give me peace, O Lord, I pray 69
Give to us eyes that we may truly see 66
Glad that I live am I 86
Gloria in excelsis Deo 7
Glory to thee, my God, this night 9
God be in my head 72
God, give me ears that will always hear thee 83
God in his love; for us lent us this planet 90
God is love; his the care 10

God is love: let heav'n adore him 11

God of concrete, God of steel 12

God save our gracious Queen 98

God who made the earth 13

God, whose farm is all creation 71

Good Christian men, rejoice and sing.. 49

Go tell it on the mountain, over the hills and ev'rywhere............................. 30

Grant us your peace, Lord... 73

Hark the glad sound! the Saviour comes 25

Heavenly Father, may thy blessing.. 75

He's got the whole wide world in his hands... 14

I danced in the morning 58

In Christ there is no East or West 74

Infant holy, infant lowly.. 31

In that land which we call holy 60

It is the joyful Easter time 50

Jesus Christ is risen to-day.. 48

Jesus, good above all other 61

Kum ba yah, my Lord, kum ba yah.. 76

Let all the world in every corner sing... 15

Life is great! So sing about it... 77

Lift up lightly the stable bar... 70

Long ago, prophets knew 27

Lord of all hopefulness, Lord of all joy.. 62

Morning has broken... 91

Now thank we all our God 16

Now tell us, gentle Mary 32

Now the green blade riseth from the buried grain.. 51

Now the holly bears a berry as white as the milk 33

O come, all ye faithful 35

O little town of Bethlehem 34

Once in royal David's city... 36

One more step along the world I go 79

O worship the King 19

Praise and thanksgiving 20
Praise to the Lord, the Almighty, the King of creation 17

Rejoice in the Lord always 18
Rejoice, the Lord is King 53
Ride on! ride on in majesty 44

See him lying on a bed of straw 37
Shalom chaverim 78
Sing a new song to the Lord 22
Sing to the Lord 21
Skipping down the pavement wide 92

Thank you, Lord, for water, soil and air 93
The angel Gabriel from heaven came 23
The holly and the ivy 26
The Lord's my shepherd, I'll not want 80
There is a green hill far away 46
There's a spirit in the air 56
The Virgin Mary had a baby boy 38
Think of a world without any flowers 94
This is a lovely world 95
Trotting, trotting through Jerusalem 45

Under Bethlem's star so bright 39
Unto us a boy is born 41
Upon the day of Pentecost 57

We plough the fields, and scatter 96
When a knight won his spurs, in the stories of old 81
When Easter to the dark world came 52
When from the sky in the splendour of summer 63
When I needed a neighbour were you there, were you there 82
When lamps are lighted in the town 97
Who was the other who died on the hill 47

Acknowledgements

The editors and publisher gratefully acknowledge permission to use the following copyright material (melody line, words and guitar chords only. Full music acknowledgement is made in the Teacher's edition):

1. Words by kind permission of J. Curwen & Sons Ltd. 2. From *Rounds and Canons* by Christopher le Fleming. Reprinted by permission of Belwin Mills Music Ltd. 4. Words from *Enlarged Songs of Praise*, melody from *Songs of Praise for Boys and Girls*, both by permission of Oxford University Press. 5. Words and melody copyright © 1972 Maranathal Music. All rights reserved. 6. Words and melody by permission of Rev. C. Micklem. 7. From *Rounds and Canons* by Christopher le Fleming. By permission of Belwin Mills Music Ltd. 8. Words reprinted by kind permission of Miss M. Egerton Smith. Melody © 1981 John Whitworth by permission of the composer. 10. Words reprinted from *Songs of Praise*, by permission of Oxford University Press. 11. Words by permission of A. R. Mowbray & Co. Ltd. Melody from the *BBC Hymn Book* by permission of Oxford University Press. 12. Words by permission of the author. Melody copyright © 1981 John Whitworth, by permission of the composer. 15. Melody by permission of the Public Trustee on behalf of the late Dr. Basil Harwood's executors. 20. Words reprinted by permission of the author. 21. Melody by permission of the composer. Words from the *Children's Mass Hymnal* by permission of Mayhew-McCrimmon Ltd., Essex. 22. Words by permission of the author. Melody reprinted by permission of the composer. 24. Words and melody © 1979 Stainer & Bell Ltd. and Methodist Church Division of Education and Youth, from *Partners in Praise*. By permission. 26. Words © 1972 Stainer & Bell Ltd., from *Partners in Praise*. By permission. 27. Words by permission of Oxford University Press. 28. Words and melody from *New Orbit*. By permission of Stainer & Bell Ltd. 31. Words from *Kingsway Carol Book* by permission of Evans Bros. Ltd. 32. Words from *Carols of the Nations* by permission of Blandford Press Ltd. 33. Words from *The Oxford Book of Carols*, by permission of Oxford University Press. 34. Melody from *The English Hymnal*, by permission of Oxford University Press. 37. Words and melody by permission of Michael Perry. 38. Melody copyright © 1945 by Boosey & Co. Ltd. Reprinted from the Edric Connor Collection by permission of Boosey and Hawkes Music Publ. Ltd. 41. Words from *Enlarged Songs of Praise*, by permission of Oxford University Press. 45. Words and melody © 1969, 1972 Galliard Ltd. By permission of Stainer & Bell Ltd. 47. Words © 1974 Stainer & Bell Ltd., from *Partners in Praise*. By permission. 49. Words by permission of Hymns Ancient and Modern Ltd. 50. Words by permission of the St. Christopher's College Trust. 51. Words from the *Oxford Book of Carols*, by permission of Oxford University Press. 52. Words from *Children Praising*, by permission of Oxford University Press. Melody © 1981 John Whitworth, by permission of the composer. 54. Words © 1979 Stainer & Bell Ltd. and Methodist Church Division of Education and Youth from *Partners in Praise*. By permission. 55. Words © 1969 Stainer & Bell Ltd. Melody © 1979 Stainer & Bell Ltd. and Methodist Church Division of Education and Youth. By permission. 56. Words by permission of Oxford University Press. 57. Words and melody © 1979 Stainer & Bell Ltd. and Methodist Church Division of Education and Youth. By permission. 58. Words and melody © 1963 Galliard Ltd. Reprinted by permission of Stainer & Bell Ltd. 59. Words by permission of Oxford University Press. 60. Words © 1979 Stainer & Bell Ltd., from *Partners in Praise*. By permission. 61. Words from *Enlarged Songs of Praise* by permission of Oxford University Press. 62. Words from *Enlarged Songs of Praise*, by permission of Oxford University Press. 63. Words by permission of Oxford University Press. Melody © 1981 John Whitworth, by permission of the composer. 65. Words and melody © 1972, 1979 Stainer & Bell Ltd., from *Partners in Praise*. By permission. 66. Words from *Someone's Singing Lord* (A. & C. Black Ltd.). Reprinted by permission of the author. Melody © 1981 John Whitworth, by permission of the composer. 67. Melody collected by Ralph Vaughan Williams, from the *English Hymnal*, by permission of Oxford University Press. 69. Words and melody by permission of Kevin Mayhew Publishers. 71. Words reprinted by permission of the author. Melody collected by Lucy Broadwood, from the *English Hymnal* by permission of Oxford University Press. 72. Melody by permission of Oxford University Press. 73. Words by permission of Kevin Mayhew Publishers. 74. Words by permission of Miss T. Dunkerley. 75. Words from *Enlarged Songs of Praise*, by permission of Oxford University Press. 77. Words and melody by permission of Oxford University Press. 78. Words from *Partners in Praise*. By permission of Stainer & Bell Ltd. 79. Words and melody © 1971, 1972 Galliard Ltd., from *New Orbit*. Reprinted by permission of Stainer & Bell Ltd. 81. Words from *Enlarged Songs of Praise*, by permission of Oxford University Press. 82. Words and melody © 1965, 1972 Galliard Ltd., from *New Orbit*. By permission of Stainer & Bell Ltd. 83. Words and guitar chords from *Music Workshop Books* (Kenneth Pont), by permission of Oxford University Press. 86. Words from *A Wayside Lute*. Melody from *Songs of Praise*, by permission of Oxford University Press. 87. Words and melody by permission of Oxford University Press. 88. Words by permission of Oxford University Press. Melody © 1979 Stainer & Bell and Methodist Church Division of Education & Youth. Reprinted from *Partners in Praise* by permission of Stainer & Bell Ltd. 89. Melody by permission of Oxford University Press. 90. Words copyright © 1973 Hymn Society of America, reprinted by permission. Melody © 1979 Stainer & Bell and Methodist Church Division of Education & Youth. Reprinted from *Partners in Praise* by permission of Stainer & Bell Ltd. 91. Words from *The Children's Bells* (OUP 1957), by permission of David Higham Associates Ltd. 92. Words and melody copyright by permission National Christian Education Council. 93. *Caring for Planet Earth* (Thank you Lord), words by permission of Oxford University Press. 94. Words and melody reprinted by permission of Stainer & Bell Ltd. 95. Words and melody © 1966 Josef Weinberger Ltd., from *Praise our Lord*. By permission of the copyright owners.

⨍ This sign means that a verse has been omitted from the original text.

The editors and publisher wish to acknowledge permission to reprint the following copyright material:

Poems

G. K. Chesterton: from *The Wild Knight* (Dent 1945). Reprinted by permission of A. P. Watt Ltd. on behalf of the Estate of G. K. Chesterton and J. M. Dent & Sons. **E. E. Cummings:** from *Complete Poems 1913-1962* (© 1956 by E. E. Cummings). Reprinted by permission of Granada Publishing Ltd. and Harcourt Brace Jovanovich, Inc. **Walter de la Mare:** from *Peacock Pie* (Faber 1969). Reprinted by permission of The Literary Trustees of Walter de la Mare and the

Society of Authors as their representative. **Eleanor Farjeon:** "Keeping Christmas" from *Puffin Quartet of Poets* (Penguin 1970). "The Quarrel" from *Morning Cockerel* (Hart-Davis 1967). Both reprinted by permission of David Higham Associates Ltd. **Carmen de Gasztold:** from *Prayers from the Ark* (1973). Reprinted by permission of Macmillan, London and Basingstoke. **Rumer Godden:** from *Prayers from the Ark* (1973). Reprinted by permission of Macmillan, London and Basingstoke. **Caryll Houslander:** from *The Morning Cockerel* (Hart-Davis 1967). Reprinted by permission of Granada Publishing Ltd. **Ted Hughes:** from *Season Songs* (1976). Reprinted by permission of Faber & Faber Ltd. **Elizabeth Jennings:** from *As Large As Alone: Recent Poems* (edited by Copeman & Gibson, Macmillan 1969). Reprinted by permission of David Higham Associates Ltd. **J. W. Johnson:** from *The Morning Cockerel* (Hart-Davis 1967). Reprinted by permission of Granada Publishing Ltd. **Lilian Moore:** from *I Feel the Same Way* © 1967 by Lilian Moore). Reprinted by permission of Atheneum Publishers. **Lew Sarett:** from *Covenant with Earth: A Selection from the Poetry of Lew Sarrett*, edited by Alma Johnson Sarett and published, 1956, by the Univ. Florida Press. Reprinted by permission of Mrs. Sarett. **James Stephens:** from *Collected Poems* (2nd. ed. 1954). Reprinted by permission of Mrs. Iris Wise and Macmillan, London and Basingstoke. **J. Walsh:** from *The Truants* (William Heinemann Ltd.).Reprinted by permission of Mrs. A. M. Walsh. **F. E. Young:** from *Morning Cockerel* (Hart-Davis 1967). Reprinted by permission of Granada Publishing Ltd.

Source for non-copyright poems
J.D. from *Poems for Assemblies* (B. H. Blackwell 1963)   **Gerard Manley Hopkins:** from *The Poems of Gerard Manley Hopkins*, edited by W. H. Gardner and N. H. Mackenzie (4th ed. 1967) published by Oxford University Press for the Society of Jesus. **Edward Thomas:** from *The Collected Poems of Edward Thomas* edited by R. G. Thomas (1978) published by Oxford University Press. **Walt Whitman:** from *Poems for Assemblies* (B. H. Blackwell 1963).

Prayers
(Where there is more than one prayer on a page the number in brackets indicates the order in which it appears):
Basil Blackwell & Mott Ltd: pp. 59; 65; 155; 161, from *Day by Day*, R. W. Purton, 1973. **Blandford Press Ltd.:** p.33(2), from *Infant Teacher's Prayer Book*, D. M. Prescott, 1964. **Victor Gollancz Ltd.:** from *God of a Hundred Names*, B. Greene & V. Gollancz, 1962. **Hodder & Stoughton Ltd.:** pp.97; 115(2), from *A Patchwork Prayer Book*, © 1976 by Janet-Lynch-Watson; p.101, from *Well God, Here We Are Again*, © 1974 by J. A. C. Bryant and D. Winter; p.33(3), from *Prayers for Children and Young People*, © 1975 by Nancy Martin. **National Christian Education Council:** pp.172; 177, from *Prayers to Use with 8-11s*, M. Putnam, 1976. **Oxford University Press:** pp.14; 53; 113; 115(1); 136, from *The Daily Service*, G. W. Briggs. **SCM Press Ltd.:** pp.13; 35; 43; 74; 169 from *Contemporary Prayers for Public Worship*, C. Micklem, 1967; pp.146(2); 167(2), from *Thinking Aloud*, John Williams, 1963. The prayers on the following pages were written for this edition at the invitation of **Leicestershire Education Committee:** pp.20; 26; 49; 63; 80; 124; 125(2); 130; 133; 138; 142; 143.

Biblical Quotations
The following quotations from the Revised Standard and Authorized Versions of the Bible are reprinted by permission of Oxford University Press.
The quotations are listed by page number and where more than one quotation appears on a page the exact source is given in brackets:
AV p.45 (Luke 1.26, 27, 30); p.51; p.56; p.97 (Acts 1.9); p.100; p.143; p.151 (Psalm 46.1); p.160 (Isaiah 35.1).
RSV p.4 (Luke 10.27); p.11 (Psalm 100.3); p.28 (Psalm 90.2); p.20; p.82; p.83; p.128; p.129; p.137 (John 14.27); p.141; p.151 (Psalm 23); p.161.

The following quotations from the New English Bible, 2nd ed. © 1970 are by permission of Oxford and Cambridge University Presses:

p.4 (Exodus 20.2, 17); p.11 (Psalm 15); p.28 (Psalm 145.9); p.87; p.94; p.97 (Matthew 28.20); p.154.

The publishers would like to thank the following for permission to reproduce photographs:

P. Almasy/World Health Organisation, p. 35; A. Anholt White, p. 9 (top); Ashmolean Museum, p. 153; Barnaby's Picture Library, p. 125; Marcus Quadros Barros, p. 155; Bodleian Library, p. 8; A. K. Boyle, p. 9 (middle); Roger Bradley, pp. 43, 107, 115; British Museum, p. 49; Brussels, Musée des Beaux Arts/Scala, p. 73; Camera Press, pp. 27, 99, 105, 143; J. Allan Cash, p. 25; Bruce Coleman/Jane Burton, p. 37; Bruce Coleman/Stephen Dalton, p. 135; Cooper-Bridgeman Library, pp. 13, 16–17, 53, 129; Courtauld Institute/Witt Library, p. 103; Escher Foundation/Haags Gemeentemuseum, p. 14; Sonia Halliday, pp. 11, 69, 87, 97; Robert Harding Associates, pp. 121, 133; Nick Hedges, pp. 145, 175; Alan Hutchison, pp. 77, 79, 149; Tonia May, pp. 21, 145; Mladinska Knija/Ivan Rabuzin, p. 157; Museum 'Het Rembrandthuis', Amsterdam, p. 83; National Gallery of Art, Washington, p. 141; Peter Newark's Western Americana, p. 167; Scala, pp. 19, 45; Nicholas Servian, FIIP/Woodmansterne Ltd., pp. 38–39; V. G. Sato, p. 127; Space Frontiers, p. 29; Ronald Sheridan, pp. 9 (bottom), 165; Warden and Fellows of Keble College, Oxford, p. 109.

Illustrations by: David Bull (aged 17), Beauchamp College, Oadby, p. 13; Dishad Talab (15), Glenfrith Hospital School, p. 57; Alex Whitworth (16), Rawlins Upper School and Community College, Quorn, p. 75; Wigston Water Leys County Infants, p. 121; Elizabeth Brothwell (17), Rawlins Upper School and Community College, p. 151; Katie Bryan (7) and Natalie Brown (7), Wigston Water Leys County Infants, p. 161; Sally Long, Leslie Marshall, Eric Tranter (all from Loughborough College of Art and Design); Alan Curless, Petula Stone.

The publishers would like to thank the Director of Education, Leicestershire, his staff and heads of Leicestershire Primary Schools for their invaluable help in advising, selecting and guiding the production of this anthology for assemblies.

Although every effort has been made to trace copyright holders, this has proved impossible in some cases. If any copyright holders incorrectly acknowledged will contact the publisher, corrections will be made in future editions.